PREFACE	7
PROLOGUE	9
1. Who Are the Basking Sharks?	13
2 Early Accounts of Basking Sharks, 1791–1900	29
3 Nuisance or Boon?	37
4 The Slaughter	47
5 Basking Sharks in British Columbia Today	61
EPILOGUE	68
APPENDIX A	71
APPENDIX B	72
APPENDIX C	74
APPENDIX D	79
APPENDIX E	80
APPENDIX F	81
APPENDIX G	82
NOTES	84
INDEX	90

PRECEDING SPREAD   *A 36-foot and a 40-foot basking shark harpooned by Einor Andersen in July of 1946 and brought to shore in Bamfield Inlet, BC. The harpoon insertion point is evident below the dorsal fin.*   PHOTO BY EINOR ANDERSEN

Basking Sharks

TRANSMONTANUS 14

Published by New Star Books
Series Editor: Terry Glavin

Transmontanus

1 A GHOST IN THE WATER *Terry Glavin*
2 CHIWID *Sage Birchwater*
3 THE GREEN SHADOW *Andrew Struthers*
4 ALL POSSIBLE WORLDS *Justine Brown*
5 HIGH SLACK *Judith Williams*
6 RED LAREDO BOOTS *Theresa Kishkan*
7 A VOICE GREAT WITHIN US *Charles Lillard with Terry Glavin*
8 GUILTY OF EVERYTHING *John Armstrong*
9 KOKANEE: THE REDFISH AND THE KOOTENAY BIOREGION *Don Gayton*
10 THE CEDAR SURF *Grant Shilling*
11 DYNAMITE STORIES *Judith Williams*
12 THE OLD RED SHIRT *Yvonne Mearns Klan*
13 MARIA MAHOI OF THE ISLANDS *Jean Barman*
15 CLAM GARDENS *Judith Williams*

Basking Sharks

THE SLAUGHTER OF BC'S GENTLE GIANTS

Scott Wallace and Brian Gisborne

TRANSMONTANUS / NEW STAR BOOKS VANCOUVER

PREFACE

This book started off as a small report on basking sharks that I prepared for the Sierra Club of Canada, British Columbia chapter. While preparing the report, I had the good fortune to be introduced to Brian Gisborne. As it turned out, Brian had already assembled hundreds of historical documents on basking sharks which were conveniently organized in binders. Upon reading some of the rich history that Brian had acquired, it was clear that the story of the basking shark needed to be presented to a wider audience, but it was also evident that there were several gaps in the historical record. Brian proceeded to diligently examine all the pertinent historical publications, government records, and narratives. I tracked down first-hand accounts of observations. We both became obsessed with finding all the historical information possibly available. The result of these esoteric pursuits is found between the covers of this book.

By coincidence, while we were assembling the information, the Committee on the Status of Endangered Wildlife in Canada posted a call for bids to write a status report on basking sharks. I was successful in this bid and was the lead author of the federal status report, collaborating with federal scientists on the Atlantic and Pacific coasts of Canada. For the federal status report, the most recent scientific information about basking sharks was assembled. *Basking Sharks: The Slaughter of BC's Gentle Giants* combines the most recent scientific knowledge along with a comprehensive historical record.

As the writer in this partnership, it is inevitable that the final text reflects my own biases and opinions more than Brian's. Nevertheless this book is the fruit of a collaboration and could not exist without the contributions of both co-authors.

*

This book would not have been possible without the generous support of Rudy North, Vicky Husband and the Sierra Club of Canada, BC Chapter. We would like to thank Terry Glavin and Betsy Nuse for their comments and edits of the original manuscript and New Star Books for all their work involved in publishing this book.

We are grateful to several people mentioned in the book who contributed images and stories. Historical research was greatly aided by George Pattern and Gordon Miller at the Pacific Biological Station library. Graeme Ellis and Pete Fletcher deserve recognition for their encouragement. Finally, on a personal note, I would like to thank my wife and family for supporting me on this esoteric pursuit.

Scott Wallace
Black Creek, Vancouver Island

PROLOGUE

It was four o'clock in the afternoon on June 3, 1791 when the scurvy-afflicted crew of the fur-trading vessel *Columbia* first viewed the northwest coast of America. Under the helm of Captain Robert Gray, the *Columbia* had left Boston on September 28, 1790 for its second expedition to the northwest coast. Its first trip had earned the vessel the reputation of being the first to circumnavigate the globe. By eight o'clock on that June evening, the *Columbia* had shortened her sails and was taking a brief rest before a 2 AM departure for what is now the west coast of Vancouver Island. At sunrise, the *Columbia* started its approach towards the landmark now called Estevan Point. John Hoskins, a purser on the vessel, noted in his logbook that "there is much wood, kelp, rockweed, whales and very large sharks" surrounding the vessel.

The mention of "very large sharks" in association with whales and floating debris suggests that the shark observed was likely the filter-feeding basking shark, whose preferred foraging areas are tidal fronts, which tend to concentrate debris. This casual record marks the earliest documented occurrence of basking sharks in Canada's Pacific waters.

For much of Canada's early recorded history, basking sharks were not given a distinctive name and were often referred to either generically as large sharks or blackfish, or misidentified with names reserved for other sharks. Even the cover photograph of this book was found by chance when searching the Royal British Columbia Museum's archives for "blackfish". Because of the confusion surrounding names, we have reassessed many early records based on descriptive accounts of the sharks' size and behaviours, photographs, season and locations. Despite problems, most of the records are unambiguous. The historical record clearly shows that this enormous fish, the size of a London bus, with a 1,000-pound liver and a brain the size of a match box, was until relatively recently a very abundant and regular inhabitant of British Columbia's coast.

That the basking shark is absent from coastal waters today reflects a shameful period of Canadian history in which we harpooned, shot, rammed, entangled and otherwise intentionally eradicated basking sharks from our Pacific waters.

This book spans a two-century history and describes the tragic story of animals once found in such high densities that they impeded the passage of a coastal steamer up the Alberni Canal in 1921. Now they are only infrequent visitors to our coast, with no more than one reported observation a year.

War Declared On B.C. Sharks

Each year for nearly two decades, the fisheries department has sent a killer expedition to the sealion rookeries of Queen Charlotte Sound waters and slaughtered sealions.

Now another venture is to be undertaken against the deepsea pirates. This time it is a private enterprise against sharks—a sort of Buccaneering raid.

Salmon canneries are planning on wiping out large numbers of the great basking sharks that frequent Rivers Inlet during the salmon seining period.

These great brutes, each weighing tons, and 30 or more feet long, have in the past caused serious damage to the nets of the thousands of fishermen who congregate there.

The sharks lie on the water and go through the nets. Their rough skins and immense strength cut and tear the seines to tatters.

This year, starting a fortnight before the season opens, it is planned to send a punitive expedition against the sharks. It will be manned by men with rifles, and it will have boats with steel-shod bows and good speed.

These boats will ram and slash the sharks. Death may come then to the sharks so attacked. The blood of the slaughtered sharks is expected to drive others away as it has done in the past.

The Province, FEBRUARY 3, 1943, P. 25

The historical record indicates that for much of the twentieth century, basking sharks were not only a common part of the flora and fauna of Canada's Pacific waters but, were so abundant that they were considered a nuisance. In 1943, at the height of World War II, another battle was brewing on the central coast in BC's salmon fishing industry. On February 3 of that year the *Province* newspaper headlined a story "War Declared on B.C. Sharks". For several years basking sharks had interfered with fishing operations near Rivers Inlet by becoming entangled in gillnets and seines, resulting in large economic losses to fishermen. In 1943, the industry began to make pre-emptive strikes on basking sharks prior to the fishing season. The *Province* headline marked the onset of a quarter-century grisly battle, endorsed by the Canadian federal Department of Fisheries, between the salmon fishing industry and basking sharks. The most cost-effective method of eradicating this massive fish proved to be a large knife on the bow of a fisheries patrol vessel which sliced the animals in half as they basked on the surface. Although the final number of basking sharks killed during this war is not possible to accurately calculate, the end result is known. Basking sharks no longer frequent our coast.

Aside from the eradication program, basking sharks also suffered tremendous losses from commercial fishing for their livers, from sport harpooning, and most of all from inadvertent entanglement with fishing gear. It is not known what remains of the population today; we do know that they are extremely rare and still falling victim to collisions with vessels and unintended capture by fishing gears.

At the time this book was written, basking sharks had not been given any special status or protection in North America. The

Committee on the Status of Endangered Wildlife in Canada will review their status in May of 2007, when we hope they will be designated as endangered.

The fossil record indicates that basking sharks have been roaming the world's oceans with much the same body shape for at least 30 million years. Recorded human interactions with the sharks in British Columbia waters date back only 214 years. During that time, basking sharks have been killed for sport, in the name of "pest control" by fisheries managers, for food, for their livers, and inadvertently by fishing gear. Furthermore, throughout this period the basking shark has masqueraded as a mythical sea serpent and as the legendary Cadborosaurus.

> **Shark Fishing Industry To Be Located On Alberni Canal**

The concerted effort to eradicate basking sharks from our coast may seem inconceivable to present-day Canadians. But in fact we continue to strip-mine many marine resources with the same disregard for ecosystem processes now as we did fifty years ago. The visuals are just more subtle. For example, over the last decade the number of deep-sea bottom trawl tows off our coast has increased tenfold. Why? Primarily for the pursuit of a little-known, deep-dwelling fish called a longspine thornyhead. These trawls damage large tracts of deep-sea benthic ecosystems. We have no idea of the potential long-term repercussions. These systems have taken millennia to develop but only decades to unravel.

The historical and ongoing story of the basking shark on British Columbia's coast is an ecological tragedy that typifies our arrogance about and ignorance of marine ecosystems. It is not only a story of eliminating a so-called pest, but also an important example of how the blind pursuit of commercial resources is often incompatible with ecosystem processes. It is a lesson about how disregard for ecosystems ultimately hurts humanity. And it is a lesson about how the media can influence our attitudes and validate abhorrent behaviours. The story of the basking shark is a reminder for us all to question our present indifference towards the complexity of marine ecosystems.

CHAPTER 1

Who Are the Basking Sharks?

> *The first clear and entire view of a basking shark is terrifying. One may speak glibly of fish twenty, thirty, forty feet long, but until one looks down upon a living adult basking shark in clear water, the figures are meaningless and without implication. The bulk appears simply unbelievable. It is not possible to think of what one is looking at as a fish. It is longer than a London bus; it does not have scales like ordinary fish; its movements are gigantic, ponderous, and unfamiliar; it seems a creature from a prehistoric world, of which the first sight is as unexpected, and in some ways as shocking as that of a dinosaur or iguanodon would be.*
>
> GAVIN MAXWELL, Harpoon Venture, 1952

Leonard Compagno, the world's foremost shark expert and author of the United Nations Food and Agriculture Organization's Sharks of the World, describes basking sharks as "supersharks" because of their large size, widespread distribution, important conservation and ecotourism values, and high media profiles. Of the 1,165 shark-like fishes of the world, only three qualify for this category, white shark and whale shark being the two others.

Despite its "supershark" status and its long history of exploitation, our knowledge of the basking shark is not much greater than it was in the 1830s. Much of our current understanding of this animal comes from scientific work done on basking sharks captured in historical commercial fisheries along the coast of Europe

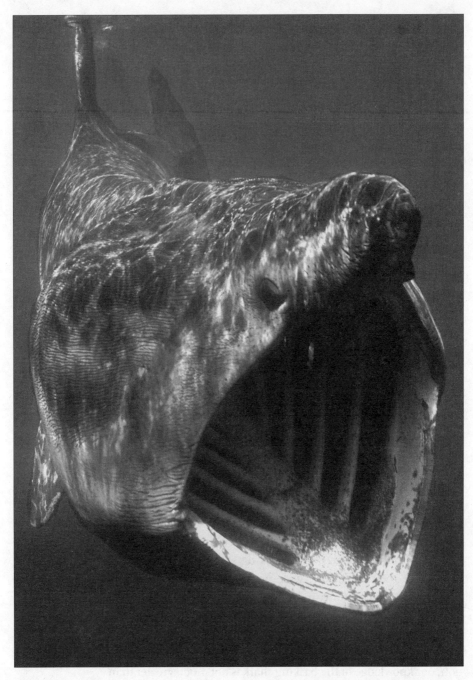

A 4.5-metre basking shark filter-feeding off the coast of Cornwall, UK, May 2005. PHOTO BY SAUL GONOR

and the United Kingdom in the mid 1900s. In recent years there have been a handful of studies in the United Kingdom and off the Atlantic seaboard of the United States using sophisticated tracking devices to understand more about their movements and habitat requirements. In Pacific Canada there has been only one study on basking sharks along with a handful of remarks recorded by scientific authorities beginning in 1891. Their life history was shaped in a prehistoric world that presented selective forces considerably different from those imposed on them by industrialized human societies, making the basking shark one of the world's most vulnerable species.

Where are they found?

Cetorhinus maximus, the basking shark, also known as the "sun shark" or "bone shark", is the second largest fish in the world (the whale shark is the largest) and can reach lengths of twelve to fifteen metres. Basking sharks seem to bask on the ocean's surface, but they are actually feeding in the plankton-rich surface waters. They are found in discrete populations throughout the temperate continental shelves of the world's oceans.

In the North Pacific, they are observed as far south and west as Japan, through to China and along the Aleutian Islands, Alaska, British Columbia, and the western seaboard of the United States and Mexico. In Canada they are found in coastal waters in both the Pacific and Atlantic oceans.

In British Columbia there are several regions where basking sharks were once known to occur in large numbers, including Queen Charlotte Sound, Clayoquot Sound, and Barkley Sound. There are dozens of other locations with confirmed records including several locations throughout the Strait of Georgia.

Movements

Although the basking shark is seemingly a large and conspicuous animal, we know very little about its movement patterns. Until recently, it was widely thought that basking sharks "hibernated" during the winter because they seemed to disappear from the sur-

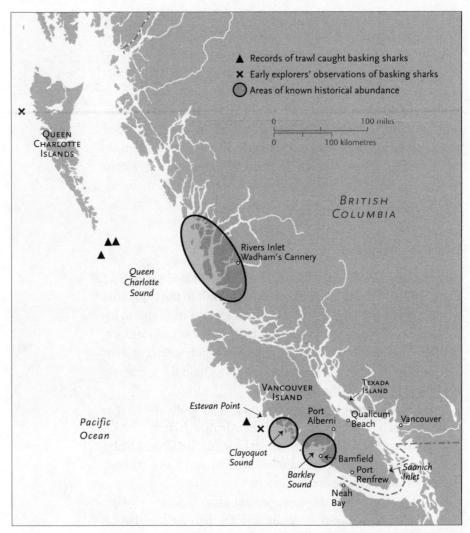

Map showing important place names mentioned in the text, areas of historically high basking shark abundance (shaded areas), present-day trawl-caught records (black triangle), and location of early explorers' records of basking sharks (cross). MAP BY ERIC LEINBERGER

face waters during these months and they shed their gill rakers, which are required for feeding.

New technologies have allowed researchers to track the depth, speed, water temperature and position of a tagged individual for up to several months. Although only nine individuals have been observed so far, these studies have not only debunked the hibernation theory, but have also shed much light on the daily and seasonal movements of these little-known animals.

There is now strong evidence that basking sharks move hundreds of kilometres to take advantage of areas with high concentrations of zooplankton. In one case, an individual tagged in the United Kingdom moved at least 3,400 km over a 162-day tracking period — an approximate maximum distance of 900 km from where it was first tracked. Off Nantucket on the US Atlantic seaboard, satellite tracking tags were placed on two individuals in close proximity to one another in September 2004. Over a period of four months, both migrated southward, one travelling approximately 1,600 km to waters off Jacksonville, Florida, and the other travelling 2,500 km to the waters between Jamaica and Haiti. Interestingly, the shark off Florida spent most of its time in the surface waters, while the shark near Jamaica spent most of its time in extremely deep waters below 480 metres.

Prior to sophisticated tracking, scientists commonly assumed that basking sharks were highly migratory because they appeared in large numbers in different localities. For example, off the Atlantic seaboard of North America they were observed in the southern portion of their range in the spring and apparently shifted to more northern latitudes by the summer. Similarly, off the Pacific seaboard they were historically observed in fall and winter in the southernmost part of their range (California) and then subsequently observed approximately 1,800 km north in waters off Washington State and British Columbia in the spring and summer. There have been no tracking studies on the west coast of North America to confirm that the pattern found on the Atlantic seaboard also occurs on the Pacific coast.

Aside from large scale seasonal movement, basking sharks are on the move on an hourly and daily basis in search of productive feeding grounds. Researchers have found that basking sharks off southwest England feed from zooplankton patches for up to twenty-seven hours before moving on. From year to year, the sharks also show broad shifts in their feeding locations, reflective of zooplankton production.

Elasmobranchs of the world and in Canada's Pacific waters

There are about 1,200 species of elasmobranchs in the world, which can be broadly categorized into two types: sharks and rays. Most species examined to date are characterized by slow growth, late maturity, and low fecundity. These attributes lead to a slow population growth rate and make them particularly vulnerable to exploitation. For many species, even a modest fishery can result in a decrease to a population over the long term. Only under carefully managed fisheries regimes could one expect to see a sustainable elasmobranch fishery. In reality, most of the world's elasmobranchs are not managed at all.

The status of elasmobranchs has come under increasing public and government scrutiny in recent years. By 2003, the Shark Specialist Group of the World Conservation Union had assessed the status of 226 species of which 62 are recognized as being threatened with extinction.

Most British Columbians are unaware that some of the world's large charismatic sharks occur in our waters, including the white, blue, thresher, and even single accounts of a hammerhead and a shortfin mako. Bluntnose sixgill sharks are commonly seen by scuba divers in select places and can reach a size of 4.8 metres (about 16 feet). There are also salmon, brown cat, Pacific sleeper, soupfin sevengill sharks, and, of course, the spiny dogfish, which is the most commonly encountered shark on our coast. It has been the focus of many fisheries over the decades. At present time, only the spiny dogfish is intentionally fished, while all of the others are to varying degrees caught as bycatch.

There are also eight species of skates: deepsea, Aleutian, sandpaper, Alaska, roughtail, broad, big, and longnose skate, and two species of rays: Pacific electric ray and the pelagic stingray. All of these species, except for the Aleutian skate, have been caught in commercial fisheries, with the primary targeted skate species being the big and longnose skates. The rest of the species are caught as bycatch.

Population structure

Through the selective forces of nature, all animals develop strategies to enhance their ability to survive. These strategies structure populations in highly ordered ways we may not completely understand. For example, resident killer whales off British

> From 1996 to 2004, about 50,000 tonnes of elasmobranchs were reported caught in British Columbia fisheries. Spiny dogfish accounts for 67.8 percent of these landings, while big, longnose, and sandpaper skates comprise 31.7 percent. The remaining 0.5 percent is made up of small but not necessarily sustainable catches of a variety of shark and skate species.
>
> All of the aforementioned species are widely distributed, ranging at a minimum from Baja California to the Gulf of Alaska. Many of these species are found throughout the North Pacific and in some cases globally.
>
> Tagging studies have confirmed the widespread distribution and migratory behaviours of many species. A long-term tagging study done off British Columbia waters found that spiny dogfish regularly undergo extensive migrations to such places as Japan, Mexico, and Alaska. Similar ocean basin migrations have been observed in blue, shortfin mako, thresher, and white sharks. Clearly, the management of many elasmobranchs requires international coordination.
>
> Elasmobranchs in general receive no special protection in Canada's Pacific waters. Sharks, excluding spiny dogfish, are not permitted to be retained in the hook-and-line fisheries but can be kept in the trawl fishery. This prevents the intentional targeting of most species. Although sharks are permitted to be retained in the trawl fishery, less than 8 percent of the non-dogfish sharks have been retained since 1996.
>
> At the present time the Committee on the Status of Endangered Wildlife in Canada is in the process of reviewing eleven species of elasmobranchs found off the British Columbia coast. These include the following sharks: basking, blue, brown cat, shortfin mako, bluntnose sixgill, soupfin, spiny dogfish, and white, as well as three species of skate: big, longnose, and sandpaper.
>
> A central conservation concern, regardless of whether a species is caught in low or high numbers, is insufficient population information. The Department of Fisheries and Oceans has recently initiated a tagging program for big skates that may provide more population data.

Columbia travel in distinct family units. Salmon return to particular streams.

Researchers do not completely understand the population structures of basking sharks. However, evidence from Scottish fisheries indicates that these animals segregate themselves based on their sex. In other words, the males and females do not ordinarily associate except during courtship and mating periods. In the Scottish fisheries the female to male ratio was 18:1. Another study of commercial catches off the United Kingdom found that sub-adult, non-pregnant females outnumbered males by 40:1. A single study in the Pacific Ocean from Japan showed a 70 percent female catch. This observed sex segregation must have relevance to the shark's conservation. If females end up the targets of surface-based fisheries, it is not surprising that basking shark populations are unable to co-exist with even small, directed fisheries.

Despite the predominance of females in areas where directed commercial fisheries have occurred, not one pregnant basking shark has been reported from commercial fisheries. Is there not only sex segregation, but also further segregation based on maturity and perhaps pregnancy? As pregnancy starts, do females segregate themselves by either depth or location from the rest of the surface "basking" individuals? Is seasonal "basking" or presence on the surface associated with sexual activity? These are some important and unanswered questions.

Diet and feeding

> *This shark's appearance, manners, and weapons do not indicate it to be a ravenous fish. One examined [stomach] contained a red pulpy mass, like bruised crabs...*
> WILLIAM YARRELL, A *History of British Fishes*, 1836

Unlike its razor-toothed relatives, the basking shark, despite its enormous size, is one of only four species of large, filter-feeding elasmobranchs found in the world. (The megamouth shark, whale shark and manta ray are the other species.) For an animal the size of a basking shark to obtain its metabolic requirements from zooplankton requires specific biological adaptations and well-defined feeding strategies.

From studies in the Atlantic, it appears that basking sharks prefer copepods, planktonic animals less than a millimetre in length. Copepods are near the base of marine food webs and are most abundant in nutrient-rich surface waters where sunlight triggers their food base through photosynthesis. This is why basking sharks are so often observed in surface waters. To capture copepods, basking sharks have evolved unique gill rakers, which  are modified skin cells on the gills. The gill rakers are similar to the teeth in a hair comb, with each tooth or raker approximately 8 cm in length. Water is strained through the gill rakers as the basking shark propels itself with its large mouth agape. Row upon row of gill rakers filter approximately a half-million litres of water per hour — the equivalent of four Olympic-sized swimming pools per day. To further maximize the surface area of the feeding structures and flow of water, basking sharks have developed enormous gill slits that extend virtually all around their head, leaving only a small amount of connective tissue at the top. Harrison Matthews, the first scientist to fully investigate basking shark biology, wondered how so "little tissue is sufficient to prevent the head falling off."

Strangely, in the winter, basking sharks shed their gill rakers. They are the only fish species known to undergo a seasonal molt of this type. At first scientists widely believed that basking sharks must become dormant during the winter because without gill rakers they are incapable of capturing small prey. Recent satellite tracking studies have provided the first evidence that this hibernation theory may not be valid. Do basking sharks switch to feeding on larger, bottom-dwelling prey or simply survive off energy stored in their massive livers? We don't know yet.

Not all areas of the ocean are able to provide enough prey for the basking shark. Given the large size of the shark and the very small size of its prey, it is not surprising that the sharks seek out the richest plankton patches. They often actively select areas where plankton concentrate — for example, fronts between colliding masses of tidal waters, geographical structures like headlands, and upwelling areas where nutrient-rich water stimulates the base of food webs. But areas of high zooplankton concentrations can change quickly over the course of a few hours, days or months, depending on the underlying process. Consequently, basking sharks seem to move around to maximize their food intake.

> ISLAND SEA MONSTER JUST 'BASKING SHARK'

Areas of high zooplankton concentrations are preferred foraging areas for other species including baleen whales and commercially valuable fishes. If basking sharks still frequented British Columbia waters, we would expect commercial fishing and tourism operations to see them. But there are virtually no recent records of basking sharks from well-monitored commercial fisheries. Indeed, the overlap of commercial fishing operations with basking shark feeding areas may possibly hinder the recovery of this depleted species.

Life history

In 1950, Harrison Matthews at the University of Bristol wrote what has proven to be the only scientific publication on basking shark reproduction. Matthews was amazed that "a fish so large, conspicuous and common as the basking shark should be practically unknown". Not much has changed in the fifty-six years since his manuscript. There is only one pregnant female on record, and juveniles have only recently been reported through various basking shark watch programs, primarily in the United Kingdom. Due to the lack of empirical evidence, some theories about basking shark reproduction, gestation, and growth have been inferred from research on closely related species.

Like other related sharks, basking sharks have internal fertilization. The developing unborn basking sharks are nourished by consuming unfertilized eggs in the uterus. This may be the only period of their lives when they use their minute teeth. The gestation period is not precisely known, but based on models and

Monster of the Deep

The three-ton basking shark is the catch of local businessmen who have been spending their Labor Day weekends the last couple of years hunting the massive creatures off the west coast of Vancouver Island. Harpoonist Art Brookman is seen at left, and George Dunn, right. They seldom land the salmon-killing monsters they hunt.

inferences from other species, it is suspected that it may be 2.6 years — the longest of any animal in the world. The litter size of basking sharks is known from only a single record where six pups were found. At birth they are estimated to be between 1.5 and 1.7 metres in length. At this large size they are not vulnerable to predation by many species except large, predatory sharks. The inter-

This appeared in the Victoria COLONIST *on September 9, 1955.*

Global basking shark fisheries

The Irish fishermen hunt it purely for the oil from its liver, wasting the rest of its huge carcase [carcass], and such an industry is in keeping with the ethics of a generation that has not scrupled to exterminate the bison for its tongue and the African elephant for its teeth.

F.G AFLALO, *British Salt-Water Fishes*, 1904

"He feels it!" was the war cry of Tex Geddes after a deep strike of a harpoon into the flesh of a basking shark. Geddes was a mate and chief harpooner aboard the converted lobster boat the *Gannet*, operating in the Hebrides under the helm of Gavin Maxwell. Maxwell's basking shark hunting operation, 1945–49, is well documented in his firsthand account *Harpoon Venture*. Because basking sharks are seemingly docile, easily approached and highly visible, they give the impression they are an easy quarry. That is, until harpooned, at which point they become an "almost unconquerable adversary". An ordinary whaling harpoon would crumple "like a corkscrew bent double, and three hundred rounds from a Breda machine gun merely pepper his flanks like pebbles."

There is a long history of commercial capturing of basking sharks from throughout the world including Norway, Ireland, Scotland, Iceland, Newfoundland, California, British Columbia, Japan, China, Peru, and Ecuador. Most of the world's historical fisheries for basking sharks (1770s–1990s) were undertaken for the valuable liver oil. The rest of the animal was typically discarded. The liver comprises 17 to 25 percent of the overall weight and yields 60 to 75 percent oil. The oil contains the compound squalene, which remains a liquid between −75 and 285 degrees Celsius, and therefore the liver oil proved useful for several industrial, cosmetic, and medicinal applications. Although the flesh has had value in some regions, it was typically discarded or at best used as fertilizer.

Modern-day reported fisheries ceased in 2003, with Norway being the last country to officially hunt the sharks. The modern-day fishery (1990s–2003)

val between litters may be every two to four years, and females are likely reproductively mature by the age of twenty. Longevity in basking sharks is unknown but could easily be fifty years. That there are no known predators of adult basking sharks suggests they have a very low rate of natural mortality.

utilized only the fins, which were exported to Asian markets. In 2001, the last year for which data is available, thirty-six sharks were killed by Norway, yielding 1,997 kg of fins with an approximate value of $2,500 per shark. The rest of the shark was dumped at sea — similar to the practice of killing buffalo for only their tongues.

The fisheries in the northeast Atlantic are relatively well documented. The first known and recorded fishery was called the Achill Island fishery off Ireland. This fishery took place annually for only a couple of weeks each year between 1770 and 1830. In the peak, up to 1,000 sharks were landed per year. The fishery is thought to have crashed due to overfishing. It was not until 1941 that large aggregations were once again observed in this region, and the fishery recommenced in 1948. It peaked in 1952 with 1,800 animals being taken, and by 1957 annual landings had fallen substantially; by the early 1960s annual landings were about fifty sharks. In 1975 the fishery closed. An astounding 12,360 sharks were taken during this period, 10,676 of them in the first ten years. Thirty years after the closure of this fishery, basking sharks are still considered to be depleted, but like several marine species, abundance trends are unknown.

Throughout the period of the Achill Island fishery, up until 2003, a much more widespread Norwegian fishery was harvesting northeast Atlantic sharks. Between 1960 and 1980 numerous sharks were taken, with an average between 1,000 and 4,000 per year. From 1970 to the present day there has been a steady decline in landings as well, fishermen exerting more effort to maintain the small catches.

The Japanese fishery has occurred to some degree for centuries. It was not until 1967 that the value of liver oil became recognized by Japanese harpooners. Between 1967 and 1978 an estimated 1,200 sharks were killed annually. By 1975 the number had dropped to 150 sharks; only about 20 sharks were taken in 1976, 9 in 1977, and just 6 in 1978. The fishery ceased completely by the 1980s, and since the 1990s a maximum of two sharks have been observed on the traditional grounds in any given year.

Susceptibility to human-caused mortality

Over millions of years, basking sharks slowly adapted to their environment by refining their body plan and behaviours to survive. What appears to be an unlikely set of adaptations has in fact been a very successful design, with the result that basking sharks are found in all of the world's oceans. Difficulties have only become apparent in the last 200 years.

Basking sharks seem completely ill-suited for survival alongside industrial human societies. Consider their traits — the sharks have a reproductive cycle on the order of five years, with perhaps the longest gestation of any animal; they do not become sexually mature until well into their teens; they spend time on the surface of near-shore waters, making them vulnerable to fisheries and vessel collisions; they segregate themselves by sex, with females being unequally vulnerable to surface threats; and, finally, their feeding areas overlap with industrial fishing fleets, making them vulnerable to entanglements and collisions. It is not surprising that human contact has rapidly depleted their populations.

The scientific record in Pacific Canada

There has been little research on basking sharks in Canada. The first Canadian scientific record comes from the Natural History Society of British Columbia in 1891. In that document it was noted that "basking sharks are plentiful in Queen Charlotte Sound during the summer months and are known to Indians who occasionally kill individuals."

It was not until 1934 that the first scientifically authenticated report of a basking shark was confirmed in correspondence between C.V. Wilby from the University of Washington and W.A. Clemens, from the Pacific Biological Station. According to this letter, this "shark is abundant in B.C. waters and is frequently seen." The first scientific description of basking sharks in British Columbia was not officially published until 1935, and at that time they were described as "common along the British Columbia coast."

There is also correspondence between Dr. Clemens and the Chief Supervisor of Fisheries, J.A. Motherwell. In April 1935, Motherwell noted while on a fur seal survey that there were "numerous humpback whales and basking sharks" approximately twenty-five miles south of Pachena on the southwest coast of Vancouver Island. In the last scientific publication of this era, the Fisheries Research Board confirmed that an eighteen-foot basking shark was found stranded near Parksville in 1943. Aside from these brief accounts there is little else written in the early scientific literature about west coast basking sharks. Later research in the 1940s described the oil and vitamin content of their livers for commercial purposes.

Clayoquot Sound is one of the few places on British Columbia's coast where basking sharks were regularly seen from the early 1970s through to the early 1990s. It was not until 1992 that whale researcher and long-time Tofino resident Jim Darling embarked on a photo identification study of basking sharks in the sound. Jim had been observing basking sharks sporadically in the area since the 1970s. Between June and August 1992, Jim and his field assistant identified twenty-seven different individuals. Two of the sharks were positively identified as males by the large white claspers in their pelvic region. The subsequent year the sharks failed to return in any numbers, and by 1994 the sharks had disappeared completely, and have yet to return. Each day during the summer there are dozens of ecotourism boat trips exploring Clayoquot Sound, along with substantial other marine traffic. If sharks were in the area today they would certainly be spotted and reported.

There are no accepted explanations for the disappearance of basking sharks from Clayoquot Sound. Does their disappearance coincide with the rapid expansion of salmon farming? In Clayoquot Sound, salmon farmers install large, netted pens perpendicular to the coastline in the areas where basking sharks were common for centuries. According to Jim Darling, basking sharks are easily entangled in anything (even a single line going to a prawn trap) so it is difficult to imagine how they could avoid the net pens of a salmon farm. Salmon farm company officials say entanglements have not occurred. Unofficial stories from employees are more equivocal. Is it a coincidence that the reduction and end of sightings of basking sharks in Clayoquot Sound correlate exactly with the introduction and increase of salmon farms in their daily range?

Unfortunately basking sharks are no longer present in Canada's Pacific waters for studies to be undertaken. Basking sharks once were common on this coast, and although there is little scientific information available to help understand their former abundance, there are alternative information sources we can use to draw out information about former habitat, seasonal distribution, and even estimates of historical population size. These sources include newspapers, interviews, photographs, and historical writings.

A sport-killed basking shark hung on display at the government dock in Bamfield, circa 1962. Note the harpoon gash below the dorsal fin. PHOTO BY BILL FULLERTON

CHAPTER 2

Early Accounts of Basking Sharks, 1791–1900

Today, several thousand people make their living operating on Canada's Pacific waters — fisherfolk, boat captains, tourism operators, and researchers. These mariners have in some cases spent decades at sea, yet when asked if they have seen a basking shark in the last thirty years, their answer is typically "no". In previous eras, things were different. Mariners made explicit reference to the abundance of basking sharks. We can only conclude from historical records that basking sharks were not rare but were part of the common flora and fauna in British Columbia's waters.

What are the chances that, within the first few hours of arriving on what is now the west coast of Vancouver Island, John Hoskins, the purser on the 1791 voyage of the fur-trading vessel *Columbia*, would encounter a basking shark? Today, this would seem improbable. Yet Hoskins's logbook entry referring to "whales and very large sharks" off what is now called Estevan Point is likely the first account of basking sharks on our coast. Sometime in July on that same voyage, he mentions that the inhabitants of the land "also get sharks which are here very large". That Hoskins called them "very large" sharks suggests that he is referring to basking sharks.

In 1794, nineteen-year-old sea captain John Boit embarked on a two-year trading voyage and circumnavigation of the globe. His vessel, the *Union*, was a 65' 5½" topsail sloop, heavily armed with cannon and a crew of twenty-two. After departing Newport in January 1794, the vessel sailed across the Atlantic to the Canary

Islands, down the coast of Africa to the Cape Verde Islands, then crossed the Atlantic again and sailed down the South American coast to the Falkland Islands. The sloop made the bumpy passage around Cape Horn and steered north, eventually sighting Cape Scott at the top end of Vancouver Island on May 16, 1795. On June 6 Boit made a brief note in his logbook of "many large whales and sharks round." The location is somewhere near the northwestern tip of the Queen Charlotte Islands. Is this also a record of basking sharks? The association with large, presumably baleen, whales lends some support to this inference.

> There is a stream in Clayoquot Sound called "Shark Creek" which is named after the Nuu-Chah-Nulth place name, mamach-aqtlnit, which is the translation for basking shark.
>
> E.M. GEORGE, 2003

The next record for "large sharks" comes from the journal of twenty-three-year-old surgeon and fur trader William Fraser Tolmie. In 1832 Tolmie left Scotland on a five-year contract with the Hudson's Bay Company. He was posted in several areas including Fort McLoughlin (Bella Bella). On Friday June 12, 1835, at Fort McLoughlin, Tolmie made an entry in his journal: "large sharks daily seen in the [McLoughlin] Bay". The size, location, date, and reference to more than one shark suggest he is referring to basking sharks. Although "large" is subjective, basking sharks are the only large sharks known to aggregate in shallow bays during the summer. In fact, there are several accounts of basking shark sightings in the month of June. Finally, the fact that Tolmie was able to observe the sharks from the surface is a good indicator that he had seen basking sharks.

William Eddy Banfield, namesake of the misspelled town of Bamfield, came to this coast aboard HMS *Constance* in 1846 and later served as an onboard carpenter. He left the service in 1849 and for several years traded among the First Nations on the west coast of Vancouver Island. His letters were regularly featured in the Victoria *Gazette*. On September 9, 1858, he reported to the *Gazette* from his location in Clayoquot Sound that "whales, dogfish and sharks are caught in large quantities here, for their oil; also, halibut, salmon and codfish." In this letter Banfield makes the distinction between dogfish and sharks, which at least rules out the small spiny dogfish as the shark species he had observed. We can't know for certain what kind of shark he is referring to, but a good piece of evidence to support the notion that Banfield is talking about basking sharks comes from the pioneering anthropologist Phillip Drucker. Drucker's book *The Northern and Central Nootkan Tribes* describes the shark fisheries of Nootkan tribes in the 1850s.

> In the 1850s or thereabouts the dogfish oil industry grew up. From then until about the end of the century many men devoted a large part of their time in late spring and summer to fishing for dogfish ... or harpooning big "mud sharks". For the big sharks the sealing harpoon, with one or two sealskin floats (of type used for whaling) on the line, were used. There seems to have been no fear of sharks nor any feeling that hunting them was dangerous or difficult. The creatures of course, were not wary and did not require such cautious stalking as seal, sea otter, or sea lion.

Drucker refers to mud sharks, which in the early 1900s was the preferred name for bluntnose sixgill sharks. Sixgill sharks are typically found in deep waters near the bottom; harpooning them would be impossible. Drucker also mentions that the shark species in question "were not wary", a behaviour characteristic of basking sharks. Finally, Clayoquot and Nootka sounds are known areas of basking shark aggregations in subsequent records. It seems likely that both Banfield's and Drucker's accounts refer to basking sharks.

Numerous Basking Sharks Menace to Nets Of Salmon Fishermen Along Coast of B.C.

In 1860 Dr. George Suckley became the only prominent naturalist at the time to suggest that basking sharks were rare. Suckley first arrived on the west coast in 1853 as a surgeon and naturalist for the American Pacific Railroad Survey between St. Paul, Minnesota and Puget Sound. He resigned in 1857 to pursue his natural history interests exclusively. In 1860 he wrote that a very large shark was captured at Port Discovery [near Cape Flattery–Strait of Juan de Fuca] in Dec. 1856. My informants told me that from its liver four barrels of oil were extracted! Large sharks are very rare in Puget Sound; so rare that it is not improbable that they are stragglers which have followed the warm "Pacific gulfstream" from more southern regions.

The size of the liver described by Suckley is characteristic of a basking shark; no other large shark would yield such large quantities of oil. Suckley's claim that large sharks are "very rare" was hotly contested in the public media in 1862 by one of the most famous chroniclers of history on the Pacific coast — James G. Swan. Swan was an oysterman, customs inspector, secretary to congressional delegate Isaac Stevens, journalist, reservation schoolteacher, lawyer, judge, school superintendent, railroad promoter, natural historian, and ethnographer who strongly disagreed with Suckley's comments.

Swan wrote an extensive, full-page article in the *Washington Standard* newspaper entitled "The Fishes of Puget Sound", in which he asserts that large sharks are actually common in Puget Sound waters.

> They [large sharks] must have swarmed here since the Doctor [Suckley] made that report, for at this time there are regular shark fisheries on different parts of the waters of the Sound; one in particular at the mouth of the Snohomish river in Possession Sound [north end of Puget Sound], and another on Hood's canal, both of which produced an amount of oil sufficient to induce several companies to prosecute the business. During the present summer while cruising in the waters of the Straits [Juan de Fuca] and along the coast both north and south say fifty miles each way, I noticed a great number of sharks, in fact so plenty were they, that instead of being as Suckley says "very rare", I have seen the water apparently alive with them particularly off San Juan or Patchina Harbour [Port Renfrew], nearly opposite Neeah Bay. Either Dr. Suckley was most grossly misinformed, or else as I before remarked, sharks have

From the PROVINCE, *July 7, 1947.*

THE ANCIENT SPORT of shark fishing by means of harpooning has been revived by Johnny Humphrey of Blubber Bay, who recently hooked more than a dozen off Texada Island. He is shown above with his home-made harpoon.

Sharkhunting Latest Sport Around Texada Island

swarmed in these waters since his observations were made. The Indians seldom capture them, for although their livers make a large amount of oil, they are difficult and dangerous customers to attack in canoes.

In 1868, Swan wrote that

> a very large species of shark, known among whalemen as "bone shark", is occasionally killed by Makahs, and its liver yields great quantities of oil. I saw one in October 1862, killed in Neeah Bay, twenty-six feet long, and its liver yielded nearly seven barrels of oil, or over two hundred gallons. These sharks are very abundant during the summer and fall, but the Indians rarely attack them except when they come in shore to feed which they do at certain times. They are easily seen by the long dorsal fin projecting above the water, and, as they appear to be quite sluggish in their movements, are readily killed with harpoons or lances. The flesh is never eaten.

From Swan's accounts, there is no doubt that basking sharks were very abundant at least around Neah Bay and likely as well along the south coast of Vancouver Island near Port Renfrew. On April 5, 1866, Swan referred in his journal to a Makah whaler who was undergoing an intense ritual, going without sleeping or eating for six days and nights and bathing in the ocean. This particular individual admitted that he could probably only persevere with this whaling ritual for two days and "if that would not suffice to make him a whaleman he could kill sharks."

Seven-Ton Shark Landed After Epic West Coast Fight

The next historical record of basking sharks is from the geologist, geographer, teacher and ethnographer George Mercer Dawson. In 1878–79 Dawson visited the Queen Charlotte Islands. His much-praised geological report on the islands is as much known for its rich account of Haida culture and natural history observations as for its geological research. On August 23,

1878, shortly after leaving Virago Sound, Dawson wrote in his logbook,

> see a very large shark, which followed the boat for some distance, occasionally showing its back fin above water. Length estimated at over 20 feet.

Based on the size, the behaviour of following boats, and the observation of the back fin above the water, Dawson is without a doubt describing a basking shark. In his 1880 published report on the Queen Charlotte Islands, he further discusses the large sharks he encountered.

> Large sharks abound on the northern and western coasts, and are much feared by the Haidas, who allege that they frequently break their canoes and eat the unfortunate occupants. No instance of this kind is known to me, but they fear to attack these creatures. When, however, one of them is stranded, or found from any cause in a moribund state, they are not slow to take advantage of its condition, and from the liver extract a large quantity of oil.

In a letter dated September 1, 1879, and addressed to Alex C. Anderson, Inspector of Fisheries for British Columbia at the time, Alexander Mackenzie, a Hudson's Bay Company representative in Masset on the Queen Charlotte Islands, describes the abundance of "sharks of large size" during the dogfish season of July, August, and September.

The final record from the 1800s is from the *Year Book of British Columbia* for 1897, in which a brief mention of the basking shark states that it is "plentiful in Queen Charlotte Sound during the summer months. It attains a great size, is perfectly harmless, and so tame that while basking it may be touched by hand."

These early records from credible sources leave little doubt that basking sharks were present in several locations on British Columbia's coastline, ranging from Puget Sound to the northern tip of the Queen Charlotte Islands, throughout the 1800s.

CHAPTER 3

Nuisance or Boon?

Beginning in the early 1900s the relationship of European settlers to basking sharks began to change. The photograph on the cover of a beached basking shark, taken in 1901, is the first record of a basking shark killed by salmon-fishing gear. During the decades to follow, countless basking sharks were killed in British Columbia, initially by accident and then by a directed, all-out slaughter. During most of the twentieth century we viewed basking sharks as we did other animals that impeded our economic progress. Thousands of seals and sea lions, mergansers, black bears, and even kingfishers were destroyed in order to "manage" fisheries.

The first written account of what might be a basking shark caught in a gillnet was published in the *Province* newspaper in 1906. The report said a "big grey shark" was caught in Toba Inlet. This story involved a thirteen-foot shark capsizing a small fishing dory and both shark and fisherman fighting for their lives, with only the fisher surviving. There is insufficient information in the article to confirm that the fish was actually a basking shark. The paper reported that the fisherman's boots were "ripped to shreds", but basking sharks have no teeth as such, so if the report is accurate, the fish could not have been a basking shark.

A December 1910 newspaper article described plans for the first industrial attempt to harvest basking sharks in British Columbia. The article stated that "sharks in abundance are found in the waters off Vancouver and Queen Charlotte Islands, numbers of

FACING PAGE: *Basking shark entangled in an illegal salmon gillnet in coastal waters near Kristiansund, Norway,* circa 1982–1985. *Several hundred basking sharks in British Columbia are thought to have been killed by similar fishing practices between 1900 and 1970.* PHOTO BY NILS AUKAN

the fish averaging twenty feet in length. The fishing will be pursued from sailing boats, the catches being brought in and treated at the whaling stations." Both the size and inferred technology for capture (i.e., harpooning) would indicate basking sharks, but the success of this venture is unknown.

The second written account of a large, unwanted shark interfering with gillnet fisheries occurred at Rivers Inlet in 1915. Although the report called the fish a "mud shark", the description (thirty-five feet long with a five-and-a-half-foot-wide mouth and weighing five tons) definitively indicates a basking shark. This article is the first undisputed account of an inadvertent shark catch, but likely speaks to what was already an ongoing occurrence. Cannery officials at the time were quoted saying "that much of the damage done to salmon nets is done by mud-sharks," implying that nets were damaged regularly.

In July 1920 the *Daily Colonist* ran a story on the superiority of shark meat to salmon, halibut, haddock and cod. In this article the basking shark was described as gregarious at certain seasons. The paper said its "fifty feet or more ... offer a wholesome, palatable and nutritious food".

In 1921 the Victoria *Times* newspaper published an article headlined "Shark Industry to be Developed on Large Scale." The reference was to a new reduction plant in Barkley Sound where "the huge basking sharks ... abound in large schools." Fishing for the basking sharks was to be carried out "after the fashion of whaling with harpoons shot from guns." In a 1921 press release announcing the organization of a new shark-fishing company, Sidney Ruck, the head of the Consolidated Whaling Company, described how "sun sharks ... race up and down [the Alberni Canal] in schools

Machine-Gunners Raid Breeding Rookeries

Another machine-gunner sends a hail of lead into the ranks of the fish raiders of the Pacific [Steller's sea lions]. The white water is stained a vivid red. A heavy toll of life is taken in this conservation effort to save the seafood wealth of the nation. Then rookery after rookery, usually situated in the heart of prolific fishing areas, is raided by the gunners.

Western Fisheries, JANUARY 1941

Blowing up harbour seals

Beginning early in the 1900s, a number of other government sponsored programs were initiated to eliminate the "enemy of salmon". One program is particularly worthy of mention.

On April 24, 1918, J. McHugh, a fisheries engineer, reported that

> the first steps were taken by the department in laying down a scheme for the destruction of the hair seals [harbour seal] which congregate in the Fraser River and which for the past few years have proved such a menace to the salmon fishing industry. It was decided on this day that a systematic search for the bars most favoured by the seals should be commenced, and after several days spent in careful examination on the sand heads at the mouth of the Fraser river, a bar was discovered which seemed to be the favoured spot of a herd consisting of, I would suggest, anywhere between two and three hundred seals.

The report proceeds to describe the technique of planting explosive mines on the haulout and then,

> at the proper moment the mines were fired, and the explosion was quite successfully accomplished. On arriving on the ground it was observed that the explosion had been more destructive than I had intended. Evidently many of the seals were lying immediately over some of the mines as their bodies were blown to atoms, not a piece larger than two inches being found.

of thousands." Their numbers were so dense that "a month ago, one of the coastal steamers ran into such a solid school of these big fellows that, packed tightly against the sides of the boat and around her bow, they stopped her completely."

How many animals were actually killed in these early basking shark fishing ventures? Fisheries statistics from the 1920s make no reference to basking sharks but list only reduction products such as fish oils, fish meal or fish fertilizer. Because many species of fish contributed to these categories it is not possible to estimate basking shark kills. Understanding the level of mortality inflicted on basking sharks during this period is critical for trying to estimate what a pre-exploitation population may have been. Did these early fisheries flop or were they highly profitable? Did only a few basking sharks get killed, or were hundreds rendered into oil?

The answers to these questions may be found someday in an untapped historical archive. Given the high value of various fish oils during this period, at a minimum we can be assured that a small-scale fishing operation for basking sharks was present in British Columbia between 1910 and 1940, possibly alongside whaling operations. Even a small ongoing fishery over a thirty-year period would have long-term consequences to the population. Throughout this same period, the interaction between fishing gear and basking sharks was likely a regular part of salmon gillnet fishing on our coast. This is suggested by a 1929 report on sea lions that briefly mentions large sharks getting entangled in gillnets in Rivers Inlet.

Not many basking shark sightings, commercial ventures or inadvertent interactions were reported in the early 1930s. Either the animals themselves were scarce, or the media did not pick up on the stories. A basking shark was on one occasion misidentified as a sea serpent, but overall the number of sharks appeared to have dropped.

Serious nuisances

Starting in 1937, fishermen in Rivers Inlet began voicing serious concerns regarding the inadvertent interference by basking sharks with salmon gillnets. By the early 1940s the problem had escalated, perhaps due to a natural shift in abundance of basking sharks or perhaps due to a change in fishing practices. But regardless of the cause, one thing became clear: the basking shark was no longer a "gentle giant" but was now described as a "great brute" and a "deep-sea pirate" that needed to be destroyed.

By 1942, stories of basking sharks tangling with the gillnets on Rivers Inlet become common, and the conflict seems to have escalated. "[H]undreds of huge basking sharks" were reported to have caused "thousands of dollars' worth of damage to gillnets in the Rivers Inlet district." The net boss for the Canadian Fishing Company at the time reported that "70 nets in a period of seven days" were lost by his boats alone.

The Rivers Inlet fishery has been the third most valuable salmon fishery in the province, behind the Fraser and Skeena Rivers, for much of the last 125 years. At its peak, thousands of fishermen worked seasonally supplying salmon to about ten canneries. During the late 1930s and early 1940s, it was estimated that there were 1,200 boats on the Rivers Inlet fishing grounds. It was a short and competitive fishing season as canneries vied for the lucrative sockeye run. Anything that impeded the catch of fish, such as the loss of fishing gear and fishing time, would drastically impact a company's profits. Hence, company officials started to take preventative and punitive measures to combat basking sharks.

By 1943, it became clear that basking sharks cruising and filter-feeding on the rich fishing grounds of Rivers Inlet could no longer be tolerated. The *Province* newspaper in February 1943 featured an article headlined "War Declared on B.C. Sharks" (see sidebar, p. 10), as did *The Fisherman*.

> **Fisherman lands 27-foot shark near Gibsons**

On July 12, 1943, the *Province* reported that the basking sharks were beginning to "show up in numbers". The mission outlined in the newspaper article was put into action. The "razor-billed shark slasher", a boat operated by BC Packers specially fitted "with a sharp steel ram", began to cut the "sleeping monsters down as they lay on the surface." This report confirmed that six sharks were "removed from life."

Meanwhile, fishermen were arranging financial measures to protect themselves from potential net damage from basking sharks in the Rivers Inlet area by establishing what the paper dubbed a "cooperative compensation scheme." Fishermen would pay annual fees into a fund that was matched by fish companies. A fisherman who suffered net damage would receive compensation from the pooled resources. The formation of such an elaborate shark insurance scheme indicates both severe and frequent encounters.

Yet after the shark-slasher press release, there were no other descriptions of the fishing grounds that year. If, by conservative estimate, only six animals were killed, did the plan work because blood drove away most of the sharks?

The following year, in June 1944, it was reported that "giant sharks are again annoying sockeye salmon fishermen at Namu" and that the sharks were "much bigger than in other years." The article gave the impression that yet another conflict between salmon fishers and basking sharks was looming, but the outcome is unknown. In July 1948 fishers reported in *The Fisherman* that "sharks are a major problem here in Smiths [Inlet] again." From that time on there were no further written accounts or sightings of basking sharks in the region.

The Rivers Inlet story needs further investigation. We don't know whether the "razor-billed shark slasher" was ever used again. A number of basking sharks were killed by entanglement in Rivers Inlet during the first half of the 1900s, but there is no way of making an accurate estimate. We do know that in the nearly sixty years since these episodes, basking sharks have not returned to Rivers Inlet.

Sea serpents, sea monsters, and Cadborosaurus

The basking shark might hold the answer to the questions surrounding the accounts of sea serpents, sea monsters, and the alleged Cadborosaurus, or "Caddy" — the name comes from Cadboro Bay near Victoria, which this creature was said to frequent. Over the years around the world numerous sightings of sea serpent-like creatures have been proven to be basking sharks. In British Columbia there were 181 "Caddy" reports between 1881 and 1991. Some of these sightings have proven to be basking sharks.

The first mystery animal that turned out to be a basking shark appeared in 1934. A skeleton found on Henry Island, south of Prince Rupert, which brought international attention to British Columbia, was rumoured to be one of the extinct Steller's sea cows or perhaps a sea monster. However, the excitement soon diminished when Dr. Clemens, director of what is now called the Pacific Biological Station, identified it as a basking shark skeleton.

In 1947 a forty-five-foot long skeleton was found near Port Alberni at Vernon Bay. There was tremendous excitement in the media that "Caddy" remains had been found, but the "hard-boiled scientists from Departure Bay puncture[d] the story like a balloon."

A year later in December 1948, a similar "Caddy-like" corpse washed up on

Shark oil boom

Concurrent with the early 1940s "pest" problem in commercial fisheries, shark oils of all species became very valuable during the heyday of war-time shark fishing. The value of shark and other fish livers was at an all-time high because of their potential for vitamin and industrial oil production, but the basking shark actually had one of the lowest vitamin A potencies of any of the west coast sharks. Fishermen only received anywhere from 3¢ to 20¢ per pound during this period (compared to 37¢ to $3.00 per pound for dogfish livers). The principal use for basking shark liver oil at that time was in the tanning process. The commercial fishery for basking sharks seems to have been small relative to some of the other oil-reduction fisheries -- in part because of the relatively low value, but mostly because of the difficulties in capturing and processing these huge animals. Despite the low value

Rathtrevor Beach near Parksville. This one was also confirmed by Dr. J.L. Hart of the Pacific Biological Station to be a basking shark, as was an unidentified "monster" that washed up in 1953 near Campbell River. The evidence of this particular basking shark stranding brought hope for reviving the thrilling shark-hunting sport that thrived in the Strait of Georgia in the 1940s.

A few skeletons washed ashore do provide some information on basking shark habitat. In particular, strandings and skeletons found in the Strait of Georgia suggest that basking sharks may have been a regular presence in the inside waters. It is also interesting that most of the stranding records were from late fall, which may have some biological significance.

The observed feeding behaviour of basking sharks gives rise to a potential explanation for sea serpents. While feeding on the surface, basking sharks often swim in tandem in groups of two or three. It is possible that a row of three sharks feeding in tandem could be mistaken for a single huge "sea serpent". Basking sharks have also been observed with their head, dorsal fin, and caudal fin simultaneously breaking the surface. The book *Cadborosaurus: Survivor from the Deep*, outlines 181 documented sightings of "Caddy" between 1881 and 1991. The authors, P.H. LeBlond and E.L. Bousfield, note a conspicuous rise in the frequency of sightings between 1930 and 1960, which coincides with the period when basking sharks were known to be most abundant.

of basking shark livers, the sheer size of an average liver (1,000 pounds plus) still made it profitable for some fishermen to harvest these animals.

Killing basking sharks was not for the feeble. The *Province* newspaper in 1944 published an article about the Huu-ay-aht whaling chief John Moses from Sarita, a small village at the mouth of the Sarita River, which flows into Barkley Sound. The whaling chief, "on any sunny day when the water is reasonably calm," would paddle his dugout canoe sometimes ten miles offshore to kill basking sharks by means of a "heavy spear attached to 600 feet of half-inch rope." Once the shark was speared, the resulting struggle would sometimes extend over a period of six hours, after which time he would tie the dead basking shark astern and set out to tow it many miles back to shore. The incentive for the trip was $80 for selling the liver, which according to Moses was "not hay."

Newspaper articles from 1946 report that "several fishboats in the Bamfield area" were using harpooning techniques in the pursuit of basking sharks. In July 1946 the *West Coast Advocate* congratulated a husband-and-wife team from Bamfield, Einor and Mona Andersen, who successfully harpooned two sharks 36 to 40 feet long. Einor Andersen, according to a resident of Bamfield at the time, was "the expert, and regularly killed lots of basking sharks, sometimes four in a single outing."

According to Mona Andersen, Einor designed the harpoon tip himself and had it machined in Port Alberni. Extending from the harpoon tip would be a twenty-foot water pipe filled with lead, which in turn was attached to a cable and then to the boat. Once Einor had successfully implanted the harpoon, Mona would throw the boat into reverse. In an interview for this book, Mona Andersen recalled that they would go out harpooning "whenever we felt like it", but that they would not get much for the livers. "We would work like hell and sell the livers to the Co-op for three or four cents per pound."

> *Coast Shark Livers Help Bombing Raids on Berlin*

In 1947 the *Province* reported that a 37-foot 10½-inch, seven-ton "granddaddy" basking shark was landed by Einor Andersen in the vicinity of Bamfield. According to the article, a passing pilot boat

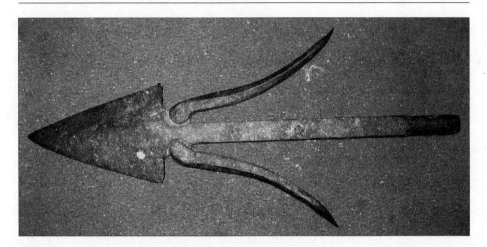

Einor Andersen's self-designed harpoon tip used for capturing basking sharks in Bamfield during the 1940s.
PHOTO BY SCOTT WALLACE

noticed a fishing vessel being towed around in circles. After hailing its skipper, the passing vessel discovered that Einor was in the process of trying to kill a "plumb nuts" basking shark, and that he had already been towed three miles. The pilot boat then came alongside to become part of the flotilla. Another hour of towing and five rifle shots to the head finally ended the shark's life.

It is impossible to know the total number of basking sharks killed by commercial harpooners. During this period, all basking shark landings were lumped into the category of "mixed shark" in federal fisheries statistics. This category was a catch-all for less valuable or infrequently landed sharks. It also included brown cat, blue, sleeper, and salmon sharks. Between 1941 and 1945 more than 800,000 pounds of mixed shark liver was harvested. But after 1945 the category of "mixed shark" disappeared completely from the government records.

It seems that although there was some economic incentive to harvest basking sharks during the early 1940s, they were more of a nuisance and cost to fishermen than they were an economic boon. At the same time as there was a small basking shark fishery in the Barkley Sound region, there were calls for eradicating the sharks. If the value of the liver was sufficiently high and basking sharks so prevalent, we would expect that a large-scale, directed fishery would have quickly emerged. Based on all the evidence, it seems likely that commercial fisheries killed, at a minimum, many hundreds of these sharks before synthetic oils and vitamins put an end to fishing sharks for their livers in 1948. With no remaining value for the basking shark liver and a seemingly uncontrollable, expanding population in the Barkley Sound area, the basking shark came to be labelled a widespread "pest" that needed to be eradicated.

Following the conflicts reported in Rivers Inlet during the early 1940s, similar tensions began to arise in Barkley Sound starting in 1945. Barkley Sound was also an extremely important salmon fishing ground at this time and appears to have supported the greatest number of basking sharks anywhere on the coast. Stories of conflict between basking sharks (called the "curse of the fisherman") and salmon fisheries started appearing in the provincial media in June of 1945. It was reported that "[i]n the Kildonan area these monsters appear to be increasing in numbers, and seine fishermen are having a lot of trouble, they are hoping the Fisheries Department will take some action to kill them off."

A month later, on July 12, 1945, the *West Coast Advocate* reported that basking sharks were playing havoc with fishermen's nets in the Alberni Canal. Several fishermen were reported to have been reducing or eliminating the "menace" by sticking them with harpoons and "blowing them up".

CHAPTER 4

The Slaughter

By 1948 gillnet fishermen were demanding a bounty on basking sharks. The reported value of basking shark livers was three cents per pound, which would mean average remuneration to a fisherman of approximately $36 to $48 per shark. However, if a shark were to get entangled in a gillnet it would typically destroy between "50 to 100 fathoms of a 200-fathom net valued at between $350 and $400." Evidently, harvesting sharks strictly for their livers was not profitable, and there were too many sharks.

In July of 1948 the Victoria *Times* reported that west coast fishermen in Barkley Sound were making an "annual cry" for a bounty on basking sharks. Shortly after the salmon fishing season began in Uchucklesit Harbour, "no less than eight of the sharks became ensnarled in nets and were dragged ashore at Kildonan cannery and killed." Each such entanglement resulted in an economic loss of about $3,000 in 2005 dollars. The article further stated that when the provincial Department of Fisheries was approached annually to initiate a bounty (like those for countless other species) for basking sharks, the department claimed basking sharks were simply a natural hazard "just like a storm or any other act of nature."

Starting in 1949, basking sharks finally made it onto the federal fisheries department's list of "Destructive Pests". Membership on the list was essentially a government-endorsed declaration for their eradication or at least control. Other varmints included black bears, merganser ducks, kingfishers, seals, and sea lions. All of these animals were regularly killed by fisheries officers while doing their nine-to-five patrols. The tallies of their patrol missions were written up in annual regional reports. Some comments from the 1963 Barkley Sound report include: "black bears were destroyed only where found on salmon streams with excessive numbers", "mergansers were killed this year in considerable numbers during the year's stream patrols", "sea lions destroyed by Departmental personnel numbered 34", and "hair seal noses submitted for bounty numbered only 20". For several years this was standard fisheries management.

For the first few years, basking sharks were noted in the annual reports but not actively eradicated. The 1949 report stated that "basking sharks appeared in Barkley Sound at the start of the sockeye season and did some damage to fishermen's nets. This year however they did not remain in the area as long as usual and damage was much lighter than it has been for the past few years." The 1950 annual report mentioned that "basking sharks appeared in large numbers during the sockeye season and did a great deal of damage to fishermen's nets."

By 1951, attitudes towards basking sharks had become increasingly antagonistic. In June of that year *The Fisherman* reported that the government was experimenting with methods to combat the sharks, presumably with harpooning techniques. The success of this first control program is unknown, but it likely did not result in many kills. The fateful year for basking sharks was 1953. All of British Columbia's newspapers reported stories of Barkley Sound basking sharks destroying the gillnets of at least thirty fishermen in early June. Although the loss was covered by shark insurance, fishermen were growing more impatient about sharing the waters with basking sharks.

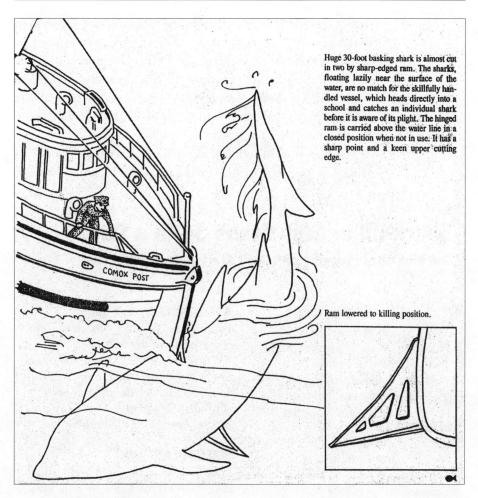

Huge 30-foot basking shark is almost cut in two by sharp-edged ram. The sharks, floating lazily near the surface of the water, are no match for the skillfully handled vessel, which heads directly into a school and catches an individual shark before it is aware of its plight. The hinged ram is carried above the water line in a closed position when not in use. It has a sharp point and a keen upper cutting edge.

Ram lowered to killing position.

A simple and lethal invention

In 1955 a more efficient, less cumbersome killing technique was invented, and basking sharks finally suffered the worst consequence of being on the "Destructive Pest" list. That year, the Department of Fisheries commissioned Alberni Engineering and Shipyards to design and install a death-dealing basking shark cutting blade on the bow of the regional fisheries patrol vessel, the *Comox Post*. The knifelike ram, although not overly sophisticated, was featured in the magazine *Popular Mechanics* in November 1956. When the crew of the *Comox Post* approached a school of basking sharks the knife would be lowered from a hinge by a cable so that the cutting edge was just below the surface of the

Illustration of the basking shark cutting blade mounted on the fisheries patrol vessel Comox Post. Published in the magazine Westcoast Fisherman in 1990.

water. The knife is now housed in the Alberni Valley Museum in Port Alberni. Seeing the knife for the first time, you are struck by its sheer simplicity: it is a sharpened, triangular chunk of steel that was likely designed in a few minutes. Paradoxically, the basking shark, which embodies eons of biological design, was incapable of evading this executioner. Sharks cruising obliviously on the surface would be sliced in half.

Newspapers went into a feeding frenzy over the shark-killing knife. On June 22, 1955, the front page of the Victoria *Times* fea-

Shark-Killer Ship Stabs 34 in a Day
Sun Man Goes Progging with Gov't Craft in West Coast Bay

By JIM HAZLEWOOD
Sun Staff Reporter

The great shark slaughter began at noon and continued for hours.

We littered the beaches with their livers and the bottom with their carcasses.

RECORD KILL

Up and down the length of Pachena Bay we sailed, slashing and rendering with the huge knife on the bow of the Comox Post.

It was a colossal fight between the ship and the sea monsters, with the ship winning all the matches.

By dusk we had a record one day kill of 34 definite, about 5 possibles and 10 misses.

This is the Federal Department of Fisheries answer to the menace of the world's biggest sharks. The marauders have begun swarming off the west coast of Vancouver Island in the past three years and we were out trying to run them down.

3 TO 5 TONS

Basking sharks range between three and five tons in size.

When they begin frisking about on the busy fishing grounds off Bamfield and Cape Beale the effect is roughly equivalent to a herd of Dinosaurs lumbering around Granville street.

BOILER PLATE KNIFE

Fisheries Inspector Bob McIndoe of Port Alberni got together with some machinists and devised the Comox Post's huge knife.

It's made of boiler-plate steel, honed to razor sharpness, and protrudes in an arc for five feet out from the bow.

We started our hunt early in the day by sailing down the Alberni canal into Trevor Channel, where sharks had been reported recently.

We spotted the first dorsal fin off Bamfield but the brute sounded before we could close on him.

An hour later another fin popped up and Captain M.O. "Red" Collette cautiously gunned the Comox Post toward it.

tured a drawing of a basking shark with the caption, "This is a basking shark, basking and leering. But the smirk will soon be wiped off its ugly face by the fisheries department, which is cutting numerous sharks down to size." The press enthusiastically described the fishermen as victims, demonized the sharks, and endorsed the eradication of this species. The press even went as far as describing the copepod-eating shark as a "salmon-killing monster". The general public was urged to help with the elimination effort by means of recreational harpooning, shooting, and ramming.

> **CLOUD OF BLOOD**
> I stood on the bow, directly above the knife.
> A torpedo shaped olive-green form made a frantic twist in the water, but it was too late.
> A cloud of blood burst over the bow as the ship shuddered at the impact. A tail that was six feet across thrashed the water until it was severed from the shark's body.
> Then the two pieces toppled off the knife and sank.
> Huge pieces of suety liver, the only part of the shark that will float, bobbed in the wake of the ship.
>
> **BAYFUL OF SHARKS**
> We felt better after that. We were worried that we might not see any sharks at all.
> Then we rounded Cape Beale into the calm confines of Pachena Bay and I knew what the primeval ocean must have looked like 1,000,000 years ago.
> The bay was literally crawling with sharks. There were dorsal fins everywhere we looked.
> Mr. McIndoe and I acted as spotters while Captain Collette jockeyed the Comox Post around the bay in an orgy of blood-letting.
>
> The sharks having no natural enemies have no fear of the boat until it is practically on top of them. Then there's a moment of confusion as they see their doom approaching.
>
> **HARPOON FAILS**
> But by the time those ponderous reflexes get working, it's usually too late.
> We slashed at them steadily for four hours.
> We tried harpooning one, but he tore off under the boat and chopped the line by pulling it against the propellor blade.
> It was grim and grisly work but vitally necessary.
> When we reported our huge kill on the fisherman's radio band, congratulations poured in.
>
> **101 THIS YEAR**
> The Comox Post started its war on the big sharks last year when [it] became evident they were increasing rapidly on the coast.
> Although the crew has to work shark-progging in with other duties, the kill of 101 to date is already higher than last year.
> *Vancouver Sun*, APRIL 24, 1956

After the initial flurry of press commentary on the shark blade in 1955 and 1956, the *Comox Post* went about its daily job, firing bullets into the occasional sea lion, seal, or merganser and slicing sharks when seasonally abundant. At the end of each fishing season an annual report was written, and over the years the entries for basking sharks appear to diminish (see Appendix G). The blade was used over a period of fourteen years in the Barkley Sound region, during which time 413 kills were recorded.

Although numerous newspaper articles describe the pest removal endeavour, one is worth transcribing in its entirety (see pp. 50–51). The article was written by a Vancouver *Sun* reporter who happened to be on the *Comox Post* for the largest recorded single-day kill of basking sharks, on April 24, 1956.

It is not surprising that in 1956 the *West Coast Advocate* newspaper featured an article headlined "Fisheries Patrol Winning War on Basking Sharks." After 1956 there are no further newspaper articles on either the commercial aspect of or the nuisance associated with basking sharks. The eradication program continued in more or less public silence for twelve more years and appears to have attained its goal of eliminating "the pest" from British Columbia's waters.

Concurrent with the *Comox Post* slicing basking sharks, other fisheries patrol vessels including the *Laurier, Howay,* and *Kitimat* were under directives from the federal Department of Fisheries to ram basking sharks if they were encountered during patrols. The sheer force of the collision was intended to mortally injure the sharks. Pete Fletcher grew up in Bamfield and eventually worked for the Canadian Coast Guard for thirty years, primarily as a lighthouse keeper and later servicing the lights. His first job at age eighteen, however, was as a deckhand on the *Laurier* in 1955. When the vessel rammed a shark, he says, it was like "going aground on a gravel bar, all the dishes would rattle." Fletcher is a self-declared animal lover. "I always loved animals, as a kid I enjoyed seeing the basking sharks. Back then basking sharks were a fact of life, as common as catching a salmon. I hated my job ... killing sea lions, seals, ramming basking sharks. I knew it was not right. I was the wrong guy in the wrong place

Shark hunters clean up on west coast pests

The basking shark cutting blade, now housed at the Alberni Valley Museum, was used by the federal Department of Fisheries' regional patrol vessel Comox Post from 1955 to 1969. PHOTO BY SCOTT WALLACE

and that's why I eventually left." Although there are no verified numbers for opportunistic killings, the former *Laurier* deckhand suggests that 200 to 300 is a good estimate.

There were also periods when the *Comox Post* was not carrying the knife and instead would use its hull to cut or ram the basking sharks. In May 1965, the logbook of the *Comox Post* recorded that there was a "good showing of basking sharks off Kelp Bay [near Bamfield], one cut by hull." It is not known how many of the hull-killed animals were reported in the annual government reports. We can infer that the overall actual number killed by all patrol vessels was conservatively reported.

Entanglements

Is it fair to blame the *Comox Post* knife for the disappearance of basking sharks from the BC coast? Not completely. The rationale for the eradication program in the first place was to decrease the nuisance factor associated with catching basking sharks in commercial salmon gillnets and, to some degree, trolling gear. When a basking shark is ensnarled in a gillnet it normally spins around in the net until it becomes completely entangled. Few sharks survived this. Once caught, some were sold commercially for their livers, but most, if not dead already, were simply killed by means of a firearm and dragged to shore in an effort to salvage what remained of the gillnet. The total number of basking sharks killed from entanglement is not known, but, given the extent of the problem, there would have been significant unreported mortality incurred by the gillnet fleet. For example, in 1952, *Western Fisheries* reported that Hugh G. Garrett, a gillnetter, caught seven basking sharks in that one season alone.

Given that basking sharks are easily caught in an assortment of fishing gear aside from gillnets, there were likely many non-reported snarl-ups with troll, crab, herring, trawl, prawn, and longline gear. These accidental entanglements would have killed an unknown number of basking sharks. Fishing interactions, over the long term, were likely the single largest source of human-caused mortality. Based on the extent of documented basking shark and gillnet interactions, we can conservatively estimate that several hundred sharks were killed in this way.

Sport fishing

The general public was also killing basking sharks. At one time, the sport of harpooning basking sharks was popular enough that the

> One day, somewhere between 1960 and 1962, I was boating up Effingham Inlet in my sixteen-foot Swish. When I got near the head of the inlet, all I could see were dorsal fins. The head of the inlet is about a half mile across. There were easily forty–fifty dorsal fins at the surface at any given time. The sharks are constantly moving, and as they swim they slowly rise and sink below the surface of the water. At any given time, only a fraction of the total numbers could actually be seen at the surface. This means there were a whole lot more there than what I could see. I was scared. What if I hit one? I decided to slowly idle out. Today, those animals would not survive because of all the boats. They would be constantly hit.
>
> — JACK GISBORNE, NOVEMBER 2004

A word about words

Both the government and the media played a role in providing misinformation on basking sharks to the public. The words that the media chose to describe basking sharks betrayed our biological understanding at the time.

Over the basking shark war years, the media and government referred to this species as *deepsea pirates, great brutes, huge ugly fish, curse of fishermen, sissies of the shark family, pests, monsters of the deep, grotesquely huge, hideous looking, menaces, marauders, fighters,* and *good-natured slobs.*

They were incorrectly portrayed as *salmon-killing monsters, huge mammals, fighters,* and *lazy.* In one instance a journalist even described them as *plankton-eating monsters,* which seems a little paradoxical.

Even the early scientists unwittingly helped set the tone with their taxonomical descriptions. In 1905 David Starr Jordan, the most renowned ichthyologist of his time, described the basking shark as a "dull and sluggish animal — almost as inert as a sawlog — clumsy and without spirit".

The words used to describe these animals in the written record helped condone our behaviours towards them.

Canadian Pacific Railway promoted so-called fishing for British Columbia's basking sharks in publicity releases in the late 1940s. Parksville was promoted as the "shark fishing mecca of the Pacific Northwest," and "scores" of anglers from California, Washington, Oregon, and even the eastern states were said to come to Vancouver Island for the sport. Apparently the sport took off after a basking shark washed up on a beach in Parksville; local anglers were inspired to catch another one. Early attempts involved snaring the sharks using "clothesline" fishing tackle. Harpooning soon caught on, with recreational fishers using 600 feet of half-inch rope with a harpoon at one end and a metal drum at the other.

As with other elements of the basking sharks' history in our waters, the sport record is incomplete. It is not possible to estimate the number of basking sharks killed for sport in the 1940s through to the mid-1960s as the only written records are in newspaper stories. The story has been partly conveyed by newspaper headlines during this period. (See Appendices C and D.)

A basking shark ensnarled in a salmon gillnet circa 1974 in Sarita Bay, Barkley Sound. Sharks accidentally captured, if not dead already, were killed in an attempt to salvage the fishing gear.
PHOTO BY HENRY GRANNEMAN. ALBERNI VALLEY MUSEUM # PN11152

Harassment

It can be assumed the newspapers only covered a small fraction of the actual encounters. For many coastal residents, harassing basking sharks was simply a way of life in the 1950s and 1960s. Eric Wickham, one-time fisherman and current president of the Canadian Sablefish Association, recounts his boyhood Bamfield pastime of ramming the sharks with small boats for fun and harpooning them for the thrill of the ride and the challenge for a "young whaler" to kill a basking shark. Pete Fletcher, who also spent his boyhood in Bamfield, remembers how basking sharks were used for target practice by staff at the Bamfield Cable Station. One shark, Old Joe, was "riddled with so many bullets that his dorsal fin looked like Swiss cheese." Old Joe, an exceptional basking shark, was a regular in Bamfield Inlet and, as such, the fishing community somehow accepted him. He apparently met his fate when a new captain on a scheduled cargo ferry thought he would do the community a favour by ramming the shark.

Harassment by harpooning, shooting and ramming was an endorsed activity in the "War on Sharks". One newspaper reported that "many Canadian and American sportsmen planned to unite forces" in the fight. Gordie Lamb, a teenager in Bamfield in the late 1950s and early 60s, participated in the eradication effort by harpooning basking sharks primarily for fun. He was successful in bringing four kills to shore with the assistance of his older friend Bill Fullerton, who was a first-year zoology student at the University of British Columbia. "I remember dissecting the brain cavity, which was about the size of a pound of butter and the actual brain was much smaller and would fit into about the size of a matchbox," Fullerton recalls.

Another fun activity for Bamfield boys at this time was using the backs of basking sharks as an Evel Knievel-style motorboat ramp. They would simply find a basking shark, open the throttle on their ten-horsepower engine and head on a course perpendicular to the shark. Upon impact, the engine would kick up, the sixteen-foot boat would take flight, and the surprised basking shark would flick its tail, causing an enormous spray of water, and dive away from the irritant, apparently unharmed.

The sport fishing and harassment record, although far from

> Oh, there were hundreds of them around, hundreds of basking sharks. You could look out in the harbour here [Bamfield] in August, and you could see two or three on any given day cruising up and down the harbour. You'd go out, and you'd see forty, fifty, sixty, just going up to Sarita and back. There were way too many, now there's way too few. So they eliminated them, but what was left over? I would say I've caught seven or eight I've had to destroy, like in my gillnet to get my net back. They wind themselves all up in it and drown themselves. So if you put 150 gillnetters in here and everyone gets one or two, that finishes off the few that are left. And I guess naturally, they move so god damn slow, they wouldn't find each other to mate anyhow.
>
> BILL MCDERMID, CITED IN PETERSON, 1999
>
> They used to destroy the nets of the fishermen so badly, there were so many of them. I mean, it was nothing to see ten or twelve of them right in the mouth of the harbour, maybe twenty or thirty in the sound at one time, you know, in sight at one time. But they put a big knife on the bow of the *Comox Post* and just cut them in half. And I guess they went a little bit too far, because now there are very few left.
>
> JOE GARCIA, CITED IN PETERSON, 1999

complete, tells us that basking sharks were obviously present in large enough numbers and sufficiently distributed to make them sport. It is particularly interesting that so many encounters occurred in the Strait of Georgia. The strait is a semi-enclosed body of water that is ecologically and geographically quite different from the rest of British Columbia's marine waters. The total number of basking sharks killed by harassment and for recreation is unknown, but may have been in the hundreds.

Coming to an end

In May 1955, a fifteen-foot shark, possibly a basking shark, was reported cruising in the area of Qualicum Beach. The following year, in August 1956, two boaters literally ran into a basking shark in Saanich Inlet. Prior to the episode, a bystander reported watching four basking sharks in the area for two hours. The following year, in November 1957, a thirty-foot basking shark was observed in Saanich Inlet. In August 1958, a "blackfish" was reported responsible for capsizing a twelve-foot boat in Oak Bay. In the same month an eighteen-foot shark confirmed to be a basking shark was caught off Bowen Island in a gillnet. A month later, in September 1958, a twenty-seven-foot, ten-inch basking shark was accidently caught by a gillnetter "fishing off a small island at the south end of Bowen Island." In July 1959, a twenty-three-

---

Sharkhunting Latest Sport Around Texada Island

BLUBBER BAY — Shark hunting is the latest sport in the waters off Texada Island. Using a home-made harpoon, Johnny Humphrey of Blubber Bay recently hooked more than a dozen good-sized basking sharks from his 24-foot fishing boat. "But you should have seen the one that got away," says Johnny. During the day's outing ten sharks were hooked with the harpoon, of which five were landed aboard.

The Province, JUNE 7, 1947, P. 5

Sharks Numerous: Harpoons New Tackle in Brentwood Waters

All you Caddy fans go into a huddle and come out with the answer. This is the kingfish story of the week at Brentwood Bay. Unrehearsed, the show put on by a school of sharks had this resort population gaping from a ringside seat all along the shoreline.

Times, JULY 5, 1952, P. 11

Giant too strong — Rampaging Sharks Best Sport For Thrill-Seeking City Dentist

Racing through the water in the wake of a harpooned, madly-threshing giant basking shark is "one of the greatest thrills in the world," according to Dr. Winston McLuhan. The hunters caught two sharks, but Dr. McLuhan said the trip wasn't as exciting as his introduction to the sport last year when they speared a three-ton monster. "That one towed our 32-foot cruiser straight out to sea for four hours," he relates.

Colonist, SEPTEMBER 9, 1955, P. 13

2-Hour Fight Lands 2,500-Pound Shark

The largest shark ever caught in Saanich Inlet—16½ feet and 2,500 pounds —was caught Wednesday on a home-made harpoon by Harry Gilbert and his wife, Mary. Their dog, Kal, went along for the ride.

Times, APRIL 20, 1956, P. 1

Return of Monsters Recalls Thrilling Sport: Island Shark Fishing To Be Revived

The big basking sharks have returned to the east coast of Vancouver Island in considerable numbers, and fishermen are planning on reviving the once-popular shark fishing sport. A Colonist reporter fishing off Ballenas island light station sighted several of the 500–1000-pound fighters and Malcom Hodgins, Qualicum Beach Boathouse fishing guide, plans to fish for them. He saw several off of Eagle Crest and one cruised in among his boats at Qualicum Beach. He fished for them several years ago. And the sport caught on.

Colonist, JUNE 5, 1956, P. 13

foot basking shark appeared in Esquimalt Harbour. These oddball sightings and mentions of basking sharks in newspapers and historical documents are useful for understanding their historical geographical extent and abundance.

Finally, in April 1962, the Smithsonian Institution in collaboration with the Vancouver Public Aquarium used the *Comox Post* to harpoon a basking shark that was used to make a plastic model replica for display. In 1964, the National Film Board of Canada produced a short documentary entitled *Shark Hunt*, which shows the *Comox Post* bringing in the basking shark used for the plastic model.

By 1970 the war unceremoniously fizzled as the enemy was conquered. The *Comox Post* no longer patrolled with a knife, children no longer pretended they were whalers, fishermen stopped complaining of basking sharks getting tangled in their gear, and mariners on the coast no longer regularly encountered basking sharks. British Columbia's post-slaughter coast is void of basking sharks. Sightings have become as rare as those of the fabled Cadborosaurus. Basking sharks have even disappeared from Clayoquot Sound, where a small, perhaps remnant, population was observed between 1970 and the early 1990s.

CHAPTER 5

Basking Sharks in British Columbia Today

Basking sharks were once extraordinarily abundant on Canada's Pacific coast. Were sharks still present today in Barkley Sound, which was historically the most important area for them, the combination of a marine biological station, a thriving ecotourism industry, ongoing commercial gillnet fisheries, a national park and a marine passenger ferry transecting their historical habitat area three times a week would certainly provide ample opportunity for sightings. But there have been none.

Unlike marine mammals, basking sharks have no known physiological requirements to be on the surface. Because of this, one might speculate that their disappearance is due to a shift to deeper habitat, out of sight of surface observers. Often the rich plankton layer can extend well below the surface, so it is within the realm of possibility that a subtle change in the environment has caused the sharks to feed below the surface. Unfortunately, available evidence suggests that this hypothesis is wishful thinking.

The basking shark is an enormous animal and therefore consumes large quantities of prey. But its prey is microscopic, so for the shark to avoid great energy expenditure in capturing its food, the prey needs to be concentrated. Looking at the ocean as a whole, it is possible to identify the areas where a basking shark can effectively achieve its metabolic requirements. Worldwide, basking sharks are most frequently observed along tidal fronts, headlands, and upwelling areas where zooplankton become concentrated. By no coincidence, these areas are also exceptionally important fishing areas; if basking sharks were in our waters we would expect interactions with fishing nets regardless of whether the sharks are visible on the surface. That is not the case.

Historically, basking sharks were observed in large numbers on the southwestern continental shelf of Vancouver Island. The southern end of the shelf is an extremely productive upwelling area; nearly half the volume of British Columbia's groundfish comes from here. In particular, Pacific hake is caught in this region in huge quantities by commercial trawl vessels. Since 1996 the trawl fleet operating throughout British Columbia has had 100 percent independent observer coverage with all species caught in the nets recorded. During the last ten years not a single basking shark has been caught. In Atlantic Canada, where basking sharks are not yet as rare, the silver hake fishery regularly captures basking sharks as bycatch. If basking sharks were still abundant in BC waters but were simply not on the surface, one would expect to find them in the observer records from the Pacific hake fishery, but such records do not mention them.

Troublesome Basking Sharks "Speared" by Patrol Vessel

Coastwide, the commercial trawl fleet has made hundreds of thousands of trawl tows throughout the continental shelf region since 1996. Despite a comprehensive observer record from this fishery, only four basking shark captures have been recorded in the last ten years. Three of the sharks were accidentally caught off the southern tip of the Queen Charlotte Islands. A photograph of a recent basking shark catch, again accidental, was taken in August 2004 in Rennel Sound, northern British Columbia.

There are no recent reports of basking sharks interacting with salmon gillnet fishing gear in British Columbia. Although commercial salmon fishing has been substantially reduced, there are still several commercial salmon fisheries operating in areas of historical interaction. Given the ease with which basking sharks become entangled in all types of fishing gear, the lack of reports suggests that the shark is simply not in our waters.

As background for this book, we informally surveyed present-day marine operators in the areas of tourism, transport, and research. Although the survey was by no means exhaustive, we contacted all the major operators in areas of historical basking shark abundance. These include Alberni Marine Transport (they operate the *Lady Rose* and *Francis Barkley* ferries around Barkley Sound), Jamie's Whaling Station (a wildlife-viewing company based for the past twenty-four years in Tofino), BC Ferries captains on a variety of routes, ecotourism operators running multi-day boat-based excursions, and marine researchers. From all of these conversations, only three confirmed basking shark sightings were reported since 1994 (see Appendix F).

The California connection

The last apparent stronghold of the sharks in BC was Clayoquot Sound. The mysterious, sudden disappearance of basking sharks from the sound leaves open a possibility that their disappearance may be due to some natural range shift to waters elsewhere. Perhaps basking sharks are simply not in British Columbia waters but instead are in adjacent US waters? The central and southern California coast was historically the other known area for basking shark aggregations in the eastern north Pacific. Reliable data shows the seasonal disappearance of basking sharks from California waters between May and July, the same time that waters off British Columbia historically experienced peak numbers, suggesting a seasonal migration. Central California is only 1,800 km from the west coast of Vancouver Island; basking sharks could cover this distance in a few weeks. If we assume that basking sharks along the Pacific coast of North America comprise one population, the status and history of California's basking sharks may have direct relevance to their status in British Columbia.

Basking Sharks, Hair Seals Still Lead West Coast Fish Fleet's Nuisance List

Within California waters, two areas in particular were known to support large numbers of basking sharks: Monterey Bay and Pismo Beach. In 1990 a California report concluded from exten-

Basking shark caught in a groundfish trawl net off northern British Columbia in August 2004.
PHOTO BY JOHN O'DRISCOLL

sive aerial survey data that basking shark populations had declined since 1970. Both Canada and California slaughtered a significant amount of the Pacific population prior to 1970. In California, a basking shark sport fishery developed in earnest in 1924. Between 1924 and 1938 an average of 25 sharks per year were killed, with up to 100 being landed in a single year. Efficiency in capturing sharks was improved when aerial surveillance was used to spot the animals. Starting in 1946 a commercial fishery targeted basking sharks for their livers. In that year, about 300 were killed, followed by approximately 200 per year between 1947 and 1949. The fishery apparently came to an end by the early 1950s. If basking sharks along the Pacific coast of North America constitute a single stock, then there is no doubt that the population would have been severely reduced in numbers in this period in both Canadian and American waters.

Despite what appears to be overwhelming evidence that the basking shark's demise was due to human activities, natural fluctuations combined with bad timing may have played a role. Basking sharks seem to disappear periodically over decades. Dr. David Starr Jordan, the most renowned ichthyologist of his time, noted in 1887 that "the basking shark is rare here, sometimes not seen for 20 years. This year [1880] several were seen in Monterey Bay." This early account provides some evidence that over a period of decades, basking sharks may naturally come and go from specific areas. They may have become abundant in British Columbia and California for some unknown natural reason beginning around 1930. If that is indeed the case, it could not have happened at a worse time for the basking shark, coinciding as it did with a time when there was little tolerance towards animals perceived to be competing with or impeding the economic efficiency of any industry. Nor was there any understanding of the sharks' life history or of the ways in which human predation could threaten their survival as a species. All things considered, humans are likely responsible for the precarious situation that basking sharks are in today.

Basking Shark School Won't Get to College

It is ironic that an animal that was targeted for eradication due to its catastrophic effect on fishing gear half a century ago, would provide economic benefits to British Columbia if present today. In other countries where basking sharks make regular appearances, a substantial wildlife-viewing industry has developed. In particular, the United Kingdom has developed tours to view basking sharks. In British Columbia, basking sharks historically congregated in some of the most important areas for nature-based tourism. The presence of these animals in areas like Barkley and Clayoquot sounds would attract even more tourists to those places.

Towards worldwide protection

Worldwide, basking sharks are considered a species threatened by extinction. Their biology and behaviour, combined with our fishing practices and increasing marine traffic in their habitat, place their populations at particular risk. In 2000, the World Conservation Union assessed basking sharks as "vulnerable" at a global level and "critically endangered" in the North Pacific. The northeast Atlantic population is currently listed as endangered under the same criteria. Since 2002, basking shark parts have come under Appendix II of the Convention on the International Trade of Endangered Species. This prohibits basking shark fins (considered an aphrodisiac and also used in traditional medicine) from entering primarily Asian markets.

Basking sharks in Canada's waters are not given any special protection, nor have they yet been recognized as a threatened or endangered species under Canada's federal *Species at Risk Act*. However, their status in Canada is being formally assessed. Given their historical presence in Canada's Pacific waters, combined with increasing knowledge of their life history, there is no doubt that the basking shark is a species at risk. In May 2007, the Committee on the Status of Endangered Wildlife in Canada will determine whether basking sharks should be given a status designation. A threatened or endangered designation would potentially allow the species to have legal protection under the *Species at Risk Act*.

Recovery of this species, even with protection under the Act, will not be immediate and is perhaps not even possible. The inherently slow growth rate of basking shark populations means that recovery will necessarily be slow even in the absence of human-caused mortality. It will be difficult to prevent the ongoing inadvertent killing of these animals, but steps could be taken throughout the west coast of North America to work towards a better understanding of both their biology and current threats.

The most important first step will be to increase reporting of sightings throughout the northeast Pacific, utilizing other marine mammal sighting networks. Since fishing operations are considered the greatest threat to basking shark populations, the indus-

> It was a calm day and we headed for Moss Landing. When we were opposite the Salinas River [Monterey Bay], as far as your eyes could see there were fins of basking sharks, ranging in size from 20 to 30 feet in length and weighing 4–7 tons apiece.
>
> SAL COLLETTO, SARDINE SEINER, 1933

try's reporting of all sightings, both on the surface or as bycatch, should be mandatory. Concerted efforts must be made to release stranded or inadvertently captured basking sharks unharmed. Scientists should collect biological information from dead animals for potential genetic and toxicology research. Finally, we all need to be more aware of the status of this shark so that people who make incidental sightings report them. A database of sightings and inadvertent captures will be necessary to plan recovery and to prevent future interference with the sharks by fishing operations.

EPILOGUE

April 24, 2006, marked the fiftieth anniversary of the single largest basking shark kill in British Columbia. The slaughter, described earlier in this book, took place in Pachena Bay near the town of Bamfield on the west coast of Vancouver Island. On that day at least thirty-four sharks were killed by the blade on the *Comox Post*. This gory anniversary passed unnoticed.

The number of basking sharks killed in British Columbia is unknown, but based on accounts and descriptions it can be estimated that several hundred sharks were intentionally slaughtered in commercial and sport fisheries and that many hundreds more were inadvertently killed as bycatch in commercial salmon fishing operations. Fisheries officers reported 419 total kills (6 from Rivers Inlet and 413 from the Barkley Sound region). An additional unreported number were killed by fisheries patrol vessels intentionally ramming them. Based on these numbers, it is likely that several thousand sharks may have been killed in British Columbia between 1920 and 1970.

How could the disappearance of an animal as conspicuous as the basking shark have happened without anyone noticing? The story I present in this book is a prime example of the "shifting baseline syndrome", a term coined by fisheries scientist Dr. Daniel Pauly. The shifting baseline refers to the incremental lowering of standards with respect to nature, in which each new generation lacks knowledge of how the environment used to be, redefines what is "natural" according to personal experience, and sets the stage for the next generation's losses.

In this book we can see how basking sharks on British Columbia's coast have been viewed along a shifting baseline. The historic record, beginning in the 1790s, shows that these animals have been observed throughout the coastal waters of British Columbia. They were "as common as catching a salmon" and were so unbelievably abundant they impeded the passage of a coastal steamer up the Alberni Canal. But our baseline has shifted, and now we see basking sharks as rare animals.

The collective mentality of the public, government, and fishermen that promoted and condoned basking shark eradication seems outlandish today. This shift in thinking provides a glimmer of hope that attitudes towards marine conservation are changing. However, one does not need to dig deep into ongoing practices in Canada to temper this optimism. It is just that today's gross mismanagement of the marine environment is usually more subtle than the blade of the *Comox Post* or beaches littered with basking shark livers. We still do not fully understand the consequences of our actions.

It is beyond the scope of this book to list the litany of current threats to marine biological diversity in British Columbia's waters. The story of the basking shark is meant to make us all think about our relationship with the ocean. If an animal as obvious as the basking shark can disappear without any public concern or discourse, what do we really know about our ocean and the thousands of little-known species that inhabit it? What are we doing to the world's oceans today that in fifty years will seem as unbelievable as our slaughter of the basking shark?

Marine ecosystems are structured by processes well beyond current human understanding, with many more unknowns than knowns. We need to establish large marine protected areas in the oceans, phase out fishing technologies that are known to cause habitat damage, move towards fisheries that are more selective in the species captured, and decrease our overall catch of all species to allow for unknown ecosystem requirements.

Because nature behaves in such an unpredictable manner, we thought while writing this book that our interpretations might turn out to be wrong, and that maybe even next year BC's coastal inlets would be inundated with basking sharks. After completing the book, we can only hope so. Only if we're wrong will most British Columbians ever have a chance to see a basking shark.

APPENDIX A

Synonyms, descriptions, and adjectives applied to basking sharks in newspapers.

NAME GIVEN	SOURCE
Sea Tiger	*Province*, July 15, 1915
Deepsea pirates	*Province*, February 3 1943
Great Brutes	*Province*, February 3 1943
Brute	*Province*, June 7 1947
Huge, ugly fish	*Times*, July 17, 1948
Curse of Fishermen	*Times*, July 17, 1948
Mr. Shark	*Times* July 5 1952
Sissy of the shark family	*Colonist*, Sept. 27, 1953
Huge mammals	*Colonist*, June 22 1955
Pests	*Vancouver Sun*, June 23, 1955
Monster of the deep	*Colonist*, September 9, 1955
Salmon-killing monsters	*Colonist*, September 9, 1955
Grotesquely huge	*Province* February 18 1956
Lethargic creatures	*Province* May 4 1956
Menace	*Colonist* May 4 1956
Mammal	*Colonist* May 4 1956
Marauders, sea monsters, menace of the world's biggest sharks	*Vancouver Sun*, May 16, 1956
Fighters	*Colonist*, June 5, 1956
Lazy, good-natured slob	*Times*, November 28 1957
Plankton eating monster	*Times*, July 17, 1959
Sleeping giants	*West Coast Fisherman*, October 1990

APPENDIX B

Locations of basking sharks in the northeast Pacific, with particular emphasis on Canadian waters, reported in media and historical documents.

LOCATION	YEAR	NUMBER OF SHARKS, COMMENTS	SOURCE
Alberni Canal	1921	Stopped vessel	Port Alberni *News*, August 31, 1921.
Astoria, Oregon	1943	1- 2000 lb liver	*The Fisherman*, August 24, 1943, p. 2.
Ballenas Island (light station)	1956		*Colonist*, June 5, 1956, p. 13.
Barkley Sound (many locations)	1943–1969	Many sharks	Many sources (see Appendix 1).
Beaver Creek Wharf, North of Nanaimo	1893	"another lot of sharks"	*Colonist*, July 30, 1916.
Bowen Island	1958	1	*Sun*, August 29, 1958, p. 29..
Brentwood Bay	1952		*Colonist*, July 5, 1952, p. 11.
	1956		*Colonist*, June 5, 1956, p. 13.
Clayoquot Sound	1973–1992		Summarized in Darling and Keogh (1994).
Cortes Island (Bliss Landing)	1942	1-1600 lb liver; dogfish net	*The Fisherman*, September 8, 1942, p. 3.
Eagle Crest, Van. Island	1956		*Colonist*, June 5, 1956, p. 13.
Esquimalt Harbour	1959	1 (23' long)	*Times*, July 17, 1959, p. 27.
Fitzhugh and QC sounds	1955		*Province*, August 13, 1955, p. 20 (magazine section).
Gibsons (small island at the south end of Bowen Island).	1958	1 (27'10")	*Sun*, September 11, 1958, p. 21.
La Perouse Banks	1935	numerous sharks	Department of Fisheries, File# 62-3-1, letter correspondence written April 15, 1935.
	1944		*Province*, June 16, 1944, p. 5.
Ladysmith	1952		*Colonist*, June 28, 1952, p. 13.
Mistaken Island	1956		*Colonist*, June 5, 1956, p. 13.
Namu (see Rivers Inlet)			
Neah Bay	1868		Swan, James G., *The Indians of Cape Flattery*, p. 29.
North Saanich (Cole Bay)	1959	1 (not confirmed)	*Colonist*, June 19, 1959, p. 21.

Appendix B

LOCATION	YEAR	NUMBER OF SHARKS, COMMENTS	SOURCE
North Saanich (Cole Bay)	1959	1 (not confirmed)	*Colonist*, June 19, 1959, p. 21.
Oak Bay	1958	Not confirmed	*Times*, August 5, 1958, p. 15.
Pachena Bay	1956	31 or 34 (single largest kill-April)	*Vancouver Sun*, May 16, 1956.
Parksville (Rathtrevor Beach)	1948	1 (skeleton) (confirmed by J.L. Hart at PBS)	*Vancouver Sun*, December 18, 1948, p. 23.
	1956		*Colonist*, June 5, 1956, p. 13.
Parksville (Arbutus Point)	1943	1 (18' long)	Fisheries Research Board of Canada Progress Report 56 (1943), p. 15.
Port Alberni	1952	1 (15' 2000 lb)	*Colonist*, July 9, 1952, p.9.
Puget Sound	1939	Gig Harbor — >1000lbs	Tacoma *Times*, October 28, 1939.
	1940	Chambers Creek	*Tacoma Natural History Bulletin*, Vol. 1, No. 6.
	1942	Commencement Bay	
	1942	Chambers Creek	
Prince Rupert (Island Point)	1937/1938	1	*West Coast Fisherman*, October 1990, p. 44–45.
Qualicum	1892	~100	*Colonist*, July 30, 1916.
	1946		*Colonist*, November 8, 1946, p. 16.
	1955	1	*Colonist*, May 31, 1955, p. 24.
	1956	1	*Colonist*, June 5, 1956, p. 13.
Queen Charlotte Sound (most likely Rivers Inlet area)	1891	plentiful	Natural History Society of BC, *The Economic Fishes of British Columbia*, Vol. 1, No. 1 (1891), pp. 20–33.
	1897		Gosnell, R. E. *Yearbook of BC*.
Rivers Inlet	1915	100s of sharks reported	*Province*, July 15, 1915, p. 3.
	1940–1948		Numerous newspapers and fishing magazines (see Appendix A)
Saanich Inlet — Tod Inlet	1956	1 hit, 4 observed	*Times*, April 20, 1956, p. 6.
	1956	1 (16.5', 2500 lbs)	*Colonist*, August 9, 1956, p. 1.
	1957		*Times*, November 28, 195,7 p. 23.
Texada Island	1947	12	*Province*, June 7, 1947, p. 5.
Uchucklesit Harbor (Barkley Sound)	1948	8	*Times*, July 17, 1948, p. 6.
Ucluelet	1946		*West Coast Advocate*, July 18, 1946, p. 14.
Ucluelet (4 miles offshore)	1955		*Colonist*, September 9, 1955, p. 13.

APPENDIX C

Headlines and titles of articles pertaining to basking sharks in the media from 1905-present day with an emphasis on British Columbia and Washington State. *Sun*:Vancouver *Sun*, *Colonist*:Victoria *Colonist*, *Times*: Victoria *Times*

HEADLINE	SOURCE
Battle with Shark	*Colonist*, September 9, 1905
Shark Killed by Millmen at Cedar Cove	*Province*, December 5, 1905, p. 1
Fisherman's Desperate Fight With Shark	*Province*, September 17, 1906, p. 16
Captured a Shark	*Colonist*, October 6, 1906, p. 8
Steamer Disabled Basking *Sun*fish (reference to an Australian incident)	*Colonist*, October 27, 1908, p. 10
Will Engage in Shark Fisheries	Unknown, December 27, 1910
Caught Shark in Salmon Net: Excitement at Rivers Inlet When Sea Tiger was Captured	*Province*, July 15, 1915, p. 3
Sees Several Sharks Among Salmon School off Mouth of Fraser	*Sun*, August 2, 1915, p. 9
Sharks in Local Waters	*Colonist*, July 30, 1916
Says Shark Meat is Very Superior Food	*Colonist*, July 25, 1920, p. 21
Shark Industry to be Developed on Large Scale	*Times*, August 25, 1921, p. 12
Shark Fishing Industry to be Located on Alberni Canal	*Port Alberni News*, August 31, 1921
Halibut Fishermen Say 'Sea Serpent' Is a Gigantic Shark	*Province*, September 27, 1925, p. 16
Mystery Beast Believed to be Ancient Seacow	*Province*, November 24, 1934, p. 1
'Sea Monster' Basking Shark Declares Fisheries Expert	*Sun*, November 26, 1934, p. 1
A Canadian 'Monster': Sea-cow, Basking Shark, or 'Cadborosaurus'?	*Illustrated London News*, December 14, 1934
A battle was seen between a twenty foot shark and six sea lions	*Western Fisheries*, March 1935, p. 20
Unidentified Sea Monster Found Dead	*Colonist*, October 8, 1936, p. 1
Gill Net Fishing Gear in Rivers Inlet	*Western Fisheries*, July 1937, p. 20
Dogfish and Sharks	*Western Fisheries*, August 1937, p. 13
Fish Meal and Oil Production and Markets	*Western Fisheries*, November 1937, p. 27
Sharks of the Seven Seas	*Western Fisheries*, February, 1938, p. 10
Sea Monster is Observed Off Oak Bay	*Colonist*, November 16, 1938, p. 5.
Fishing Fleet News: Seattle Seiner Lands 10-ton Shark	*Western Fisheries*, December 1938, p.14
Narrow Escape	*Zeballos Miner*, January 23, 1939
Report of Gillnet Fishing in Sooke Area During 1938	*The Fisherman*, May 9, 1939, p. 1

Appendix C

HEADLINE	SOURCE
Sharks in B.C. Waters But Harmless	*Pacific Coast News*, July 27, 1939, p. 1
Tittle Tattle	*The Fisherman*, September 12, 1939, p. 8
Reports from Fishing Grounds	*The Fisherman*, July 30, 1940, p. 1
Here and There (Smiths Inlet)	*The Fisherman*, July 22, 1941, p. 4
Here and There (Smiths Inlet)	*The Fisherman*, July 29, 1941, p. 4
Fishing Ground Reports	*The Fisherman*, July 14, 1942, p. 2
Sharks Take Profit: Fishermen Report Their Nets Ruined	*Province*, August 14, 1942, p. 28
Here and There (Bliss Landing)	*The Fisherman*, September 8, 1942, p. 3
War Declared on B.C. Sharks	*Province*, February 3, 1943, p. 5
War Against Sharks is Declared	*The Fisherman*, February 9, 1943, p. 4
Riflemen to Hunt Sharks	*Sun*, March 20, 1943, p. 23
Coast Shark Livers Help Bombing Raids on Berlin	*Sun*, March 30 1943, p. 13
Fishermen Plan Mutual Aid Against Sharks	*Sun*, July 8, 1943, p. 19
Six Sharks Killed by Packers Boat	*Province*, July 12, 1943, p. 25
Fishing Ground Reports (Good Hope Cannery, Rivers Inlet)	*The Fisherman*, July 20, 1943, p. 3
How Insurance for Nets Works	*The Fisherman*, July 20, 1943, p. 2
Net Insurance Popular at Margaret Bay	*The Fisherman*, August 17, 1943, p. 3
Shark Liver Weighs 2,000 Lbs. (Astoria, Oregon)	*The Fisherman*, August 24, 1943, p. 2
Three-ton Shark Yielded 1,155 Lbs. of Liver	*Commercial Fishermen's Weekly*, August 27, 1943, p. 318
Paddles Out To Sea To Fight Sharks	*Province*, June 10, 1944, p. 5
Sharks Tear Salmon Nets Near Namu	*Province*, June 27, 1944, p. 23
Giant Sharks Spoil Catch	*Colonist*, June 29, 1944.
Fishmen Say Giant Sharks Tearing Nets	*News Herald*, June 29, 1944, p. 1
Damage By Sharks	*Colonist*, July 1, 1944, p. 21
Net Insurance at Margaret Bay	*The Fisherman*, July 18, 1944, p. 3
Giant Shark Loses Liver In Fight With Fisherman	*Sun*, May 25, 1945, p. 17
Catch 29-Ft. Shark; Liver Weighs 1747 Lbs.	*Province*, June 4, 1945, p. 22
Fish Shorts (Alberni Canal)	*The Fisherman*, June 5, 1945, p. 5
Sharks Destroy Nets	*West Coast Advocate*, June 28, 1945, p. 1
Sharks	*West Coast Advocate*, July 5, 1945, p. 11
Results Follow Shark Editorial	*West Coast Advocate*, July 12, 1945, p. 1
Basking Sharks Menace Fishing on West Coast	*Province*, July 16, 1945, p. 21
Removal of Shark Menace to Salmon Fishing is Urged	*West Coast Advocate*, August 2, 1945, p. 11
B.C. Fishermen Battle Shark Peril	*Province*, November 17, 1945, magazine, p. 8
Bamfield News	*West Coast Advocate*, July 11, 1946, p. 5
Shark Harpooned, Caught Off Nanaimo	*Sun*, July 11, 1946, p. 5
Shark's Liver Brings $66	*West Coast Advocate*, July 11, 1946, p. 10

HEADLINE	SOURCE
Harpoon, Rifle 'Catch' Sharks off B.C as Livers Sought for Vitamin Oil	*Fisheries News Bulletin*, Vol. 17, No. 203–4
Sleeping Fisherman Nets Big Mud Shark	*Colonist*, November 8, 1946, p. 16
Harpoon Favored In Shark Fishing	*Sun* December 3, 1946, p. 21
Sharking	*British Columbia Digest*, March 1947, p. 73–74
Sharkhunting Latest Sport Around Texada Island	*Province*, June 7, 1947, p. 5
Sharks Damage Salmon Nets As Rivers Inlet Run Opens	*Province*, July 2, 1947, p. 33
Numerous Basking Sharks Menace to Nets Of Salmon Fishermen Along Coast of B.C.	*Colonist*, July 16, 1947, p. 17
Seiners Get Big Haul	*The Fisherman*, August 8, 1947, p. 2
Seven-Ton Shark Landed After Epic West Coast Fight	*Province*, September 30, 1947, p. 30
Sharks Long as Street Cars Romp Off B.C. Coast	*Sun* October 11, 1947, p. 5
Scientists Say Monster Not Caddy	*Times*, December 8, 1947, p. 3
'Caddy' Remains May Be Only Basking Shark	*Sun*, December 8, 1947, p. 1
Identifies Skeleton of Enormous Basking Shark	*Colonist*, December 13, 1947, p. 22
Island Sea Monster Just 'Basking Shark'	*News Herald*, December 13, 1947, p. 4
Skeleton is of Record Size Basking Shark	*West Coast Advocate*, December 18, 1947, p. 1
British Columbia's Basking Shark	*Province*, December 29, 1947, p. 4
Log of the Mistral	*The Fisherman*, July 9, 1948, p. 6
Bounty on Basking Sharks Wanted by Coast Fishermen.	*Times*, July 17, 1948, p. 6
Log of the Mistral	*The Fisherman*, July 23, 1948, p. 6
Basking Sharks in Puget Sound and Washington Coastal Waters	*Tacoma Natural History Bulletin*, Vol. 1, No. 6
Hunting Sharks Done With Harpoons, Rifles	*Colonist*, December 12, 1948, p. 17
Maybe 'Caddy' Mystery Now Solved	*Sun*, December 18, 1948, p. 23
War's Shark Liver 'Bonanza' Over As Industry Almost Dead	*Times*, May 9, 1950, p. 2
There Were Ten (Poem)	*Western Fisheries*, December, 1950, p. 25
Basking Mud Sharks Mistaken for 'Caddy'	*Sun*, May 31, 1951, p. 11
Special Gear Tried Against Sharks in Barkley Sound	*The Fisherman*, June 19, 1951, p. 2
Shark's Fin Bill's Soup Not Serpent	*Times*, September 10, 1951, p. 13
Big Shark Observed in Ladysmith Harbor	*Colonist*, June 28, 1952, p. 13
Sharks Numerous: Harpoons New Tackle in Brentwood Waters	*Colonist*, July 5, 1952, p. 11
Shark Big, But Basker, Not Biter	*Colonist*, July 9, 1952, p. 9
Gill-Netting and Basking Sharks	*Western Fisheries*, November 1952, p. 12
Sharks, Seals Bring Misery to Fishermen	*Colonist*, June 13, 1953, p. 13
Sharks, Hair Seals Harry BC Fishermen	*Province*, June 14, 1953, p. 5
Basking Sharks, Hair Seals Still Lead West Coast Fish Fleet's Nuisance List	*Colonist*, June 23, 1953, p. 2
Seals, Sharks Raid B.C. Fishers' Nets	*Sun*, June 24, 1953, p. 8
Shark in Gillnet Proves Costly to Q.C. Fisherman	*Western Fisheries*, September 1953, p. 22
Doubt Raised That Monster Really Shark	*Colonist*, September 20, 1953, p. 17
Sharks Return May Revive Thrilling Sport	*Colonist*, September 27, 1953, p. 5

Appendix C

HEADLINE	SOURCE
Clever Marksman	*West Coast Advocate*, May 26, 1955, p. 1
Cruising Shark Sighted Off Qualicum Beach	*Colonist*, May 31, 1955 p. 24
B.C. Sharks 'Cut Up' By Unique Bow Ram	*Colonist*, June 22, 1955, p. 1
New Weapon Destroys Basking Sharks	*Province*, June 23, 1955, p. 7
Ship Fights Sharks Off B.C. Coast	*Sun*, June 23, 1955
Find New Method of Despatching the Basking Shark	*West Coast Advocate*, June 23, 1955, p. 1
Troublesome Basking Sharks 'Speared' by Patrol Vessel	*Western Fisheries*, July 1955, p. 19
Sliced Basking Shark Newest Fishery Dish	*The Fisherman*, July 5, 1955
Knife-like Ram On Vessel's Prow Kills Sharks	*Trade News*, August 1955, Vol. 8, No. 2, p. 5
B.C.'s big basking sharks are giants with no bite	*Province*, August 13, 1955, magazine, p. 20
Giant too strong ... Rampaging Sharks Best Sport For Thrill-Seeking City Dentist	*Colonist*, September 9, 1955, p. 13
Basking Sharks Face New Device	*Province*, February 18, 1956, p. 2
2-Hour Fight Lands 2,500-Pound Shark	*Times*, April 20, 1956, p. 6
Owner Seeks Home For Ton of Shark	*Colonist*, April 20, 1956, p. 1
Fisheries Patrol Winning War on Basking Sharks	*West Coast Advocate*, April 26, Sec. II, p. 4
Boat Kills 31 Sharks in One Day	*Times*, May 3, 1956, p. 1
Fishermen Ready Boats and Gear for Season Catch	*West Coast Advocate*, May 3, 1956, p. 1
Shark Hunt	*Trade News*, May 1956, p. 13
Shark hunters clean up on west coast pests	*Province*, May 4, 1956
Basking-Shark Killer Rids Coast of Menace	*Colonist*, May 4, 1956, p. 20
Steel Prow Halts Invasions of Sharks	*Sun*, May 7, 1956, p. 11
Basking Shark School Won't Get to College	*The Fisherman*, May 8, 1956, p. 3
Shark Killer Ship Stabs 34 in a Day	*Sun*, May 16, 1956
Week-End Holiday — For Some	*Colonist*, May 27, 1956, p. 2
Fewer Basking Sharks	*West Coast Advocate*, May 31, 1956, p. 5
Island Shark Fishing to Be Revived	*Colonist*, June 5, 1956, p. 13
Shark Shakes Boaters	*Colonist*, August 9, 1956, p. 1
Ship Spears Shark	*Popular Mechanics*, November 1956, p. 172
Fishermen Tell Whale of a Tale But Who Likes Shark Shenanigans?	*Times*, November 28, 1957, p. 23
Frisky Fish Blamed As Boat Overturned	*Times*, August 5, 1958, p. 15
18-Foot Shark Caught In Net Off Bowen	*Sun*, August 29, 1958, p. 29
2nd Shark Captured	*Sun*, September 11, 1958, p. 21
Fisherman lands 27-foot shark near Gibsons	*Province*, September 11, 1958
Shark Panics Swimmers Children Dash to Shore	*Colonist*, June 19, 1959, p. 21
Relax: Those sharks aren't man-eaters!	*Province*, June 20, 1959, p. 1
Sharks Abound, But No Man-Eaters Here	*Times*, July 8, 1959, p. 4
23-Foot Shark in Harbor	*Times*, July 17, 1959, p. 27
Big Shark Tours Harbor	*Colonist* July 18, 1959, p. 13

HEADLINE	SOURCE
Hunting the Giant Basking Shark	*Western Fisheries*, August 1960, pp. 19–20, 36–37
Shark sought in canal for aquarium	*Province*, April 5, 1962
B.C. shark to grace U.S. institute	*Province*, April 24, 1962
Tow-ton Shark Copied in Barkley Sound Last Summer	*Tofino–Ucluelet Press*, March 14, 1963, p. 7
Nothing to Fear From Sharks Around Here	*Colonist*, June 27, 1975, p. 39
Forty-foot Sharks Cruise Around Vancouver Island	*Islander*, November 1975
From the Goodlad Album (Photograph and caption)	*West Coast Fisherman*, April 1990, p. 61
1. Sleeping Giants 2. Ship Spears Sharks	*West Coast Fisherman*, Oct. 1990, p. 44–49
Huge knife sliced basking shark	*Islander*, September 1992

APPENDIX D

Record of basking shark photos in media.

DATE	SOURCE	PICTURE DESCRIPTION/ CAPTION	SPECIES
Sept. 3, 1938	*Province*, p. 33	Dead Marauder	Sixgill or maybe basking
August 14, 1942	*Province*, p. 28	Raiders Caught	Basking shark
June 7, 1947	Vancouver Daily *Province*, p. 5	The Ancient sport of shark fishing by means of harpooning (Picture of a man with a spear)	Reference to Basking shark
July 6, 1952	*Colonist*	Home made barb sinks home	
April 20, 1956	*Times*, p. 6	Harry Gilbert and Shark	Basking shark
April 20, 1956	*Colonist*, p. 1	Largest shark ever caught in Saanich Inlet	Basking shark
May 27, 1956	*Colonist* (magazine section), p. 2	22' basking shark	Basking shark
August 13, 1955	*Province* (magazine section), p. 20	Here's one big fellow stranded on a beach	Basking shark
Sept. 9, 1955	*Colonist*, p. 13	Monster of the Deep	3-ton basking shark
August 1960	*Western Fisheries*, p. 18	Hunting the Giant Basking Shark	
April 1990	*West Coast Fisherman*, p. 61	From the Goodlad Album	Basking shark on shore

APPENDIX E

Specific locations of basking sharks in Barkley Sound from newspapers, the *Comox Post* logbooks, and other anecdotal accounts. *Comox Post* logbooks available only for the years 1963, 1965–1967, and 1970; found in Alberni Valley Museum, Port Alberni.

DATE	# OF KILLS / OBSERVATIONS	LOCATION	SOURCE
June 23, 1955	100s	Tide rips off Cape Beale	*West Coast Advocate*, June 23, 1955
April 24, 1956	31/Hundreds	Pachena Bay	*Victoria Times*, May 3, 1956
circa 1960–1962	100	Effingham Inlet	John (Jack) Gisborne, personal comment
July 5, 1963	7	Between Sarita and Bamfield	*Comox Post* logbook
July 6, 1963	1	Between Sarita and Bamfield	*Comox Post* logbook
July 11, 1963	1	Sandford Island	*Comox Post* logbook
July 26, 1963	6	Kelp Bay	*Comox Post* logbook
August 25, 1963	1	San Mateo Bay	*Comox Post* logbook
August 28, 1963	2	San Mateo Bay	*Comox Post* logbook
May 17, 1965	7	Trevor Channel	*Comox Post* logbook
May 19, 1965	1	Sarita Bay	*Comox Post* logbook
June 8, 1967	1	Whistle Buoy	*Comox Post* logbook

APPENDIX F

Present-day sightings of basking sharks in British Columbia.

DATE	NAME/AFFILIATION	LOCATION	COMMENTS
1973	Gisborne, B. (Juan de Fuca express water taxi)	Head of Bamfield Inlet, Barkley Sound	
1979	Stewart, Anne (Bamfield Marine Sciences Centre)	Trevor Channel, Barkley Sound	
1982	Stewart, Anne (Bamfield Marine Sciences Centre)	Trevor Channel, Barkley Sound	35' long
1984	Watson, Jane (Malaspina University College)	Trevor Channel, Barkley Sound	Present for a week
1994	See paper by Darling and Keogh (1994)	Clayoquot Sound sightings from 1973–1992	
1999	Mitchell, Jim (Canada Department of Fisheries and Oceans)	48 39 50N, 124 50.8W	12', in 8 m of water
2002	Kattler, D. (BC Ferries 2nd Officer)	30 mi SW of Rose Spit (53 43.1 131 18.95)	July
2005	Lloyd, Kitty (Blue Water Adventures)	Queen Charlotte Islands	July

APPENDIX G

Recorded basking shark kills and comments under the heading "Destructive Pests". From annual federal fisheries reports titled: "Barkley Sound Area: Brief history by years covering salmon fishing activity, closures applied, escapement, and general remarks for 1949-1969." Note that reports from 1953 and 1954 were not found.

YEAR	QUOTES FROM REPORTS	KILLED
1949	"Basking sharks appeared in Barkley Sound at the start of the sockeye season and did some damage to fishermen's nets. This year however they did not remain in the area as long as usual and damage was much lighter than it has been for the past few years."	
1950	"...[A]ppeared in large numbers during the sockeye season and did a great deal of damage to fishermen's nets."	
1952	"Basking sharks did not appear so numerous in Barkley Sound this year and consequently damage to Sockeye gill-nets was not too serious."	
1955	"Predators as usual inflicted their toll on fish and fishermen, with the basking shark again in the limelight. These sharks appeared in Barkley Sound in late February and remained a menace to gill-nets until June, at which time the bulk of them moved offshore where they hampered trolling operations. After rather futilely attempting to reduce their numbers by harpooning, permission was granted by the Department to have a knife-like weapon installed on the bow of the patrol vessel. This device, after a few strengthening modifications, proved very effective and a total of 65 sharks were killed during the year, evoking many favorable comments from fishermen."	65
1956	"Basking sharks were again present in large numbers in and off Barkley Sound. By use of the shark knife mounted on the FPC [Fisheries Patrol Cutter] 'Comox Post', 105 were destroyed, following which very few reports of net damage were received by fishermen."	105
1957	"Were again present in and off Barkley Sound in quite large numbers, although evidently decreased from the previous year judging by the lighter net damage. Only 7 were destroyed due to the fact that the boat was in refit during the time the sharks were most prevalent."	7
1958	"Were again present in quite large numbers but did not show on the surface very often during the hot summer.... [T]he 'Comox Post' destroyed a total of 52.... Considerable net damage was caused by the sharks during October, and during the sockeye fishing in summer."	52

Year	Report	Count
1959	"Were as usual present in quite large numbers in Barkley Sound during the Spring, Summer, and Fall, and they were destroyed by means of the knife mounted on the FPC 'Comox Post' whenever seen.... Considerable damage to salmon gillnets were reported throughout the season, mainly during the summer Sockeye fishery in Alberni Inlet."	47
1960	"Basking Sharks: Were again very numerous in Alberni Inlet and Barkley Sound, causing considerable damage to gillnets. However, as they did not often show on the surface, only eleven were destroyed"	11
1961	"...[A]s usual very numerous in Barkley Sound but, except for May and part of June, they did not often show, whle still doing considerable damage to salmon gillnets."	32
1962		20
1963		37
1964	"Were quite numerous in Barkley Sound during the summer, but none were destroyed due to the absence of the refit of the FPC 'Comox Post', which is the only vessel adapted to carry the shark knife."	0
1965		8
1966	"The destruction of basking sharks in the Barkley Sound subdistrict this year was nil. Although the FPC 'Comox Post' has the knife located at Ecoole for quick attachment there were no basking sharks reported. For some reason this past year they were not showing at the surface."	0
1967	"21 basking sharks were destroyed in Barkley Sound by the Departmental personnel using the Comox Post Shark Knife attachment. Three nets were destroyed by these fish in 1967. One was a total loss, and the other two were 60% losses. 21 sharks were destroyed in two days.... No damage to nets was reported after that date, and sightings of the animals decreased considerably."	21
1968	"Trollers and one gillnetter reported basking sharks tangling up and destroying their gear on May 17th. The shark knife was installed and eight basking sharks were destroyed on May 22nd. The sharks then moved out of the area."	8
1969	"No control program was carried out on the basking shark population. Six reports were received of nets being damaged and two nets were completely destroyed. One shark was strangled in a gillnet. The shark knife was installed on the FPC 'Comox Post' with 'nil' results."	0

NOTES

PROLOGUE

Howay, F.W., ed. *Voyages of the Columbia to the Northwest Coast 1787–1790 and 1790–1793*. Portland: Oregon Historical Society Press / Massachusetts Historical Society, 1990.

Maxwell, Gavin. *Harpoon Venture*. New York: Viking Press, 1952.

Schnute, J., R. Haigh, B. Krishka, A. Sinclair, and P. Starr. "The British Columbia longspine thornyhead fishery: Analysis of survey and commercial data (1996–2003)." Ottawa: Canadian Science Advisory Secretariat, Research Document 059, 2004.

CHAPTER 1

Aflalo, F.G. *British Salt-Water Fishes*. London: Hutchinson, 1904.

Bailey, B.E. "Marine oils with particular reference to those in Canada," *Fisheries Research Board of Canada Bulletin* No. 89, 1952.

"Basking Shark, *Cetorhinus maximus*," *Fisheries Research Board of Canada Progress Report* 56, 1943, p. 15.

"Biology of the basking shark (*Cetorhinus maximus*)," http://www.elasmo-research.org/education/shark_profiles/cetorhinus.htm/

Clemens, W.A., and G.V. Wilby. "The sharks of British Columbia waters," *Fisheries Research Board of Canada Progress Report* 23, 1935, pp. 3–6.

Clemens/Wilby correspondence, Department of Fisheries and Oceans Archives, Box #B32 (Sharks), File 36–16, Vols. 1 and 2.

Compagno, L.J.V. *Sharks of the World. An Annotated and Illustrated Catalogue of Shark Species Known to Date*, Vol. 2, *Bullhead, Mackerel and Carpet Sharks (Heterodontiformes, Lamniformes and Orectolobiformes). Fisheries and Oceans Species Catalogue for Fishery Purposes*. Rome: Fisheries and Oceans, 2001.

Compagno, L.J.V. "Sharks, fisheries, and biodiversity." Honolulu: Paper presented at Shark Conference, February 21, 2000.

Darling, J.D., and K.E. Keogh. "Observations of basking sharks, *Cetorhinus maximus*, in Clayoquot Sound, British Columbia," *Canadian Field-Naturalist* 108, No. 2, 1994, pp. 199–210.
Darling, Jim, interviews, November 2003 and July 2004.
Department of Fisheries, File #62–3–1, April 15, 1935.
Gosnell, R.E. *Year Book of British Columbia*, Victoria: Government of British Columbia, 1897.
Green, A. "The economic fishes of British Columbia." *Natural History Society of British Columbia* 1, No. 1, 1891, pp. 20–33.
Matthews, L. H. "The Shark That Hibernates," *New Scientist* 280, 1952, pp. 756–759.
Maxwell, *Harpoon Venture*.
Pauly, D. "Growth and mortality of the basking shark *Cetorhinus maximus* and their implications for management of whale sharks *Rhincodon typus*," in F.A. Dipper, S.L. Fowler, and T.M. Reed, eds., *Elasmobranch Biodiversity, Conservation and Management*. Proceedings of the International Seminar and Workshop, Sabah, Malaysia, July 1997. International Union for Conservation of Nature, Shark Specialist Group, Switzerland, 2002, pp. 199–208.
Priede, I.G. "A Basking Shark (*Cetorhinus maximus*) Tracked by Satellite Together With Simultaneous Remote Sensing." *Fisheries Research* 2 (1984), pp. 201–216.
Sims, D.W. "Threshold foraging behaviour of basking sharks on zooplankton: Life on an energetic knife-edge?" *Proceedings of the Royal Society Biological Sciences* 266, 1999, pp. 1437–1443. An Olympic swimming pool's volume is 3.125 million litres.
Sims, D.W., and V.A. Quayle. "Selective foraging behaviour of basking sharks on zooplankton in a small-scale front," *Nature* 393, 1998, pp. 460–464.
Sims, D.W., E.J. Southall, A.J. Richardson, P.C. Reid, and J.D. Metcalfe. "Seasonal movements and behaviour of basking sharks from archival tagging: No evidence of winter hibernation," *Marine Ecology Progress Series* 248, 2003, pp. 187–196.
Skomal, G.B. "Basking Shark Tagging Update," Massachusetts Division of Marine Fisheries *News* 25, 2005, p. 6.
Skomal, G.B., G. Wood, and N. Caloyianis. "Archival Tagging of a Basking Shark, *Cetorhinus maximus*, in the Western North Atlantic," *Journal of the Marine Biological Association* 84, No. 4619, 2004, pp. 1–6.
Swain, L.A. "The Pacific coast dogfish and shark liver oil industry," *Fisheries Research Board of Canada Progress Report* 58, 1944, pp. 3–7.
Swain, L.A., and B.H. McKercher. "Examination of the unsaponifiable matter of marine animal oils 3," *Fisheries Research Board of Canada Progress Report* 65, 1944, pp. 67–69.
United Kingdom CITES Proposal. "Inclusion of the basking shark (*Cetorhinus maximus*) on Appendix II of CITES. Conference to the Parties," Proposal 12.3, 2002.
Watkins, A. *The Sea My Hunting Ground*. London: Heinemann, 1958.
Yarrell, W. *A History of British Fishes*. London: John Van Voorst, 1836.

CHAPTER 2

Boit, J. *Log of the* Union: *John Boit's Remarkable Voyage to the Northwest Coast and Around the World, 1794–1796*. Edmund Hayes, ed. Portland: Oregon Historical Society, 1986. See entry for June 6, 1795.

Dawson, G.M. *The journals of G.M. Dawson, 1875–1878*. Vancouver: University of British Columbia Press, 1989, p. 510. Both the behaviour and size indicate that Dawson is referring to a basking shark.

Dawson, G.M. "Geological Survey of Canada, Report of Progress, 1878–9: Report on the Queen Charlotte Islands," Montreal: Dawson Brothers, 1880. Reference to an encounter Dawson had with a shark on August 19, 1878 while with Indian "Jim".

Doig, I. *Winter Brothers: A Season at the Edge of America*. London: Harcourt, Brace, Jovanovich, 1980, p. 120.

Drucker, P. "The Northern and Central Nootkan Tribes," Smithsonian Institution Bureau of American Ethnology *Bulletin* 144. Washington: U.S. Government Printing Office, 1951.

George, E.M. *Living on the Edge: Nuu-chah-nulth History From an Ahousat Chief's Perspective*. Winlaw, BC: Sono Nis Press, 2003.

Goode, G.B. *The Fisheries and Fishing Industries of the United States: Section 1, Natural History of Useful Aquatic Animals*. Washington: U.S. Government Printing Office, 1886, p. 668.

Howay, F.W., ed. *Voyages of the* Columbia *to the Northwest Coast 1787–1790 and 1790–1793*. Portland: Oregon Historical Society Press / Massachusetts Historical Society, 1941, pp. 181–182, 206, 284.

Suckley, G. "All fish exclusive of the salmonidae," in *Reports of Explorations and Surveys to Ascertain the Most Practicable and Economical Route for a Rail Road from the Mississippi River to the Pacific Ocean made under the Direction of the Secretary of War in 1853–5*, vol. 12, book 2, Washington: Thomas H. Ford Printer, 1860, p. 367.

Swan, J.G. *The Indians of Cape Flattery*. Smithsonian Institution, 1868.

Tolmie, W. F. *The Journals of William Fraser Tolmie: Physician and Fur Trader*. H.T. Mitchell, ed. Vancouver, 1835.

Victoria Gazette, September 9, 1858, no. 7.

CHAPTER 3

Andersen, Mona, interview with author, Port Alberni, July 22 and November 5, 2004.

Bailey, B.E. *Marine oils with particular reference to those in Canada*, Ottawa: Fisheries Research Board of Canada Bulletin No. 89, 1952.

"Caught shark in salmon net: Excitement at Rivers Inlet When Sea Tiger was Captured." *Vancouver Province*, July 15, 1915, p. 3.

"Coast fishermen report large shark catches," *West Coast Advocate*, July 18, 1946, p. 14.

"Dogfish and Sharks," *Western Fisheries*, August 1937, p. 13.

"Fisherman's desperate fight with shark," *Vancouver Province*, September 17, 1906, p. 17.

"Fishermen say giant sharks tearing nets," *Victoria Colonist*, June 29, 1944, p. 8.

Forester, J.E., and A.D. Forester. *British Columbia's Commercial Fishing History*. Surrey, BC: Hancock House, 1975.
Fletcher, Pete, telephone interview, July 14, 2004.
"Gill net fishing gear in Rivers Inlet," *Western Fisheries*, July 1937, p. 20.
"Harpoon favored in shark fishing," *Vancouver Sun*, December 3, 1946, p. 21.
"Log of the *Mistral*", *The Fisherman*, July 9, 1948, p. 6.
"Paddles out to sea to fight sharks," *Vancouver Province*, June 10, 1944, magazine section, p. 5.
"Razor-billed shark slasher," *Canadian Fisherman*, July 1943, p. 25.
"Results follow shark editorial," *West Coast Advocate*, July 12, 1945, p. 1.
"Rivers Inlet. Wadham's Cannery showing blackfish caught by fisherman," BC Archives collections, Visual Records, No. D-02035. Forrester and Forrester (1975) have described the event behind this photograph as a "forty-foot blackfish caught in a salmon net".
"Seven-ton shark landed after epic west coast fight", *Vancouver Province*, September 30, 1947.
"Shark fishing industry to be located on Alberni Canal," *Port Alberni News*, August 31, 1921.
"Shark industry to be developed on large scale." *Victoria Times*, August 25, 1921, p. 12.
"Sharks destroy net," *West Coast Advocate*, June 28, 1945, p. 1.
"Shark's liver brings $66," *West Coast Advocate*, July 11, 1946, p. 5.
"Sharks take profit: Fishermen report their nets ruined," *Vancouver Province*, August 14, 1942, p. 28.
"Sharks tear salmon nets near Namu," *Vancouver Province*, June 27, 1944, p. 23.
"Six sharks killed by packers boat," *Vancouver Province*, July 12, 1943, p. 25.
Wailes, G.H., and W.A. Newcombe. "Sea lions," Art Historical and Scientific Association of Vancouver, *Museum and Art Notes* 4, No. 2, 1929.
"War declared on B.C. sharks," *Vancouver Province*, February 3, 1943, p. 25.
"Will engage in shark fisheries." Publication unknown, December 27, 1910, page number unknown.

CHAPTER 4

"18-foot shark caught in net off Bowen," *Vancouver Sun*, August 29, 1958, p. 29.
"23-foot shark in harbor," *Victoria Times*, July 17, 1959, p. 7.
"BC shark to grace US institute," *Vancouver Province*, April 24, 1962, p. 2.
"BC sharks 'cut up' by unique bow ram," *Victoria Times*, June 22, 1955, p. 1.
"Bounty on basking sharks wanted by coast fishermen," *Victoria Times*, July 17, 1948, p. 6.
"British Columbia's basking shark," *Vancouver Province*, December 29, 1947, p. 4.
"Clever marksman," *West Coast Advocate*, May 26, 1955.
Comox Post logbooks, Alberni Valley Museum. "Cruising shark sighted off Qualicum Beach," *Victoria Colonist*, May 31, 1955, p. 24.
"Doubt raised that monster really shark," *Victoria Colonist*, September 20, 1953, p. 17.

"Fisheries patrol winning war on basking sharks," *West Coast Advocate*, April 26, 1956, Sec. II, p. 4.

"Fisherman lands 27-foot shark near Gibsons," *Vancouver Province*, September 11, 1958, p. 23.

"Fishermen tell whale of a tale but who likes shark shenanigans?" *Victoria Times*, November 28, 1957, p. 23.

Fletcher, Pete. Telephone interview, July 14, 2004.

"Frisky fish blamed as boat overturned," *Victoria Times*, August 5, 1958, p. 15.

Fullerton, Bill. Telephone interview, April 14, 2005.

"Gill-netting and basking sharks," *Western Fisheries*, November, 1952, p. 12.

Inflation calculator: http://www.westegg.com/inflation/

Jordan, D.S. *A Guide to the Study of Fishes*, New York: Henry Holt, 1905.

Lamb, Gordon. Telephone interview, April 12, 2005.

LeBlond, P.H., and E.L. Bousfield. *Cadborosaurus: Survivor From the Deep*, Saltspring Island: Horsdal and Schubart Publishers, 1995.

Peterson, J. *Journeys: Down the Alberni Canal to Barkley Sound*. Lantzville: Oolichan Books, 1995.

Scholey, Mary. Interview.

"'Sea monster' basking shark, declares fisheries expert," *Vancouver Sun*, November 26, 1934.

"Shark shakes boaters," *Victoria Colonist*, August 9, 1956, p. 1.

"Sharks' return may revive thrilling sport," *Victoria Daily Colonist*, September 27, 1953.

"Ship spears shark," *Popular Mechanics*, November 1956, p. 172.

"Special gear tried against sharks in Barkley Sound," *The Fisherman*, June 19, 1951, p. 2.

Wickham, E. *Dead fish and fat cats: A no nonsense journey through our dysfunctional fishing industry*. Vancouver: Granville Island Publishing, 2002.

CHAPTER 5

Darling, J.D., and K.E. Keogh. "Observations of basking sharks, *Cetorhinus maximus*, in Clayoquot Sound, British Columbia." *Canadian Field-Naturalist* 108 (2), 1994, pp. 199–210.

Fowler, S.L. "*Cetorhinus maximus* (North Pacific subpopulation)," IUCN Red List of Threatened Species, 2004 (http://www.redlist.org). Downloaded on December 12, 2005.

Lien, J. and L. Fawcett, "Distribution of Basking Sharks *Cetorhinus maximus* incidentally caught in inshore fishing gear in Newfoundland," *Canadian Field-Naturalist* 100, 1986, pp. 246–252.

PacHarvTrawl database, 1996-2004.

Pauly, D. 1995. "Anecdotes and the shifting baseline syndrome of fisheries," *Trends in Ecology and Evolution* 10 (10), 1995, p. 430.

Squire, J.L. "Observations of Basking Sharks and Great White Sharks in Monterey Bay, 1948–1950," *Copeia* 1967 (1), pp. 47–54.

Squire, J.L. "Distribution and Apparent Abundance of the Basking Shark, *Cetorhinus maximus*, Off the Central and Southern California Coast, 1962–1985," *Marine Fisheries Review* 52 (2), 1990, pp. 8–11.

Tomas, T. "Dancing On the Belly of the Shark and Other Adventures on Monterey Bay," *Quarterly Bulletin of the Monterey History and Art Association* 53, 2004, pp. 3–25.

Wallace, Scott. "Status report on basking shark based on International Observer Program from 1970–2004." International Observer Program database, 1970–2004, Fisheries and Oceans Canada, Atlantic Region.

INDEX

Achill Island fishery 25
Aflalo, F.G. 24
Alberni Canal 10, 38, 46, 50
Alberni Engineering and Shipyards 49
Alberni Marine Transport 62
Andersen, Einor 6, 44–45
Andersen, Mona 44
Anderson, Alex C. 35
aquaculture. *See* salmon farming.
Atlantic Canada: basking shark sightings/captures 15, 62

BC Packers 41
Ballenas Island 59
Bamfield 30–31, 44, 57, 58
Banfield, William Eddy 30
Barkley Sound 15, 38, 44, 54, 56: basking shark locations 80; basking sharks since 1970 51; commercial basking shark fishery 38; war on basking sharks 45–53
basking shark cutting blade 49–53
basking sharks: basking 15, 20; behaviour 15, 17–20, 22–24, 61–63, 65; commercial uses 24–25, 32, 34, 43, 66; description 10, 13, 57; diet and foraging areas 9, 15, 17, 20–21, 61–62; endangered status 19, 66; fierce fighter 24, 44–45, 59; first documented in BC 9; and fishing nets 11, 27, 36, 53–54, 56, 58, 62; gill rakers 20–21; harassment of 55, 57–58; hibernation 15, 21; historical records 29–35, 72–73; misidentified 9–10, 12, 40, 42–43; other names for 9–10, 15, 31; photos 5, 14, 23, 28, 36, 56, 64; population on BC coast 10–11, 15, 25–27, 29, 61–65; range 15–17; range in BC 15–16, 35, 72–73, 80; scientific knowledge about 12, 13–20, 25–27; slaughter of 51–55, 57–59; steps to recovery 66–67; vulnerability to humans 14–15, 22–24, 65. *See also* British Columbia basking sharks, California basking sharks

BC Department of Fisheries 47
BC Ferries 63
Bella Bella 30
black bears 37, 48
Boit, John 29–30
Bowen Island 58
Brentwood Bay 59
British Columbia basking sharks: numbers killed 11, 68; numbers killed commercially 38–39, 45; numbers killed by entanglement 42, 54; numbers killed for fisheries management 52–53, 68; numbers killed by sport fishing and harassment 55, 58; population 39, 54, 57, 69; range 15, 16–17 (map), 35; sightings since 1970s 60, 61–63, 81
British Columbia: elasmobranchs 18
Brookman, Art 23

Cadborosaurus 12, 42–43
California basking sharks 17, 63, 64, 66
Campbell River 43
Canada: research on basking

Index

sharks 25–27
Canadian Department of Fisheries: basking shark kills 68, 82–83; develops cutting blade 49–50; endorses war on basking sharks 10, 11; list of "Destructive Pests" 47–48; reports mentioning basking sharks 82–83; ships ordered to ram sharks 52; study of skates 19
Canadian Fishing Company 40
Canadian Pacific Railway: advertises basking shark fishing 54
Canadian Sablefish Association 57
Clayoquot Sound 15, 30–31: basking shark research 26; disappearance of basking sharks 27, 60, 63; Shark Creek 30
Clemens, W.A. 26
Collette, M.O. "Red" 50–51
Colletto, Sal 66
Committee on the Status of Endangered Wildlife in Canada (COSEWIC) 11, 19, 66
Comox Post 49, 50–53, 59–60
Compagno, Leonard 13
Consolidated Whaling Company 38
Convention on the International Trade of Endangered Species 66
copepods 20, 21

Darling, Jim 26–27
Dawson, George Mercer 34–35
dogfish 18–19: captured for oil 31, 43
Drucker, Phillip 31
Dunn, George 23

elasmobranchs 18: and commercial fishing 18; endangered status 19; filter-feeders 20
Esquimalt Harbour 58
Estevan Point 9

First Nations: and basking sharks 26, 34. *See also specific nation*
fisheries management 12, 18–19, 37, 38, 39, 47–48, 69–70
fishing industry: basking shark bycatch 18, 37, 38, 64; basking shark net entanglements 27, 36, 53–54, 56, 58; bounty on basking sharks 47; commercial capture of basking sharks 11, 24–25, 37–40, 43–45, 65; cost of nets 47; elasmobranches 18–19; overlaps basking shark foraging areas 21, 62; problems with basking sharks 10–11, 37, 38, 40–42, 46–48, 58; reports of basking sharks since 1970s 52; wages war on basking sharks 10–11, 41, 50–53
Fletcher, Pete 52, 57
Fort McLoughlin 30 (ALSO Bella Bella)
Fullerton, Bill 57

Garcia, Joe 58
Garrett, Hugh G. 54
Geddes, Tex 24
George, E.M. 30
Gilbert, Harry 59
Gilbert, Mary 59
Gisborne, Jack 54
Gray, Robert 9

Haida nation 34–35
hair seal. *See* harbour seal
harbour seal 37, 39
harpooning 48, 51, 57: basking shark response to 24, 44–45; commercial fishing 24, 44–45; First Nations fishing 31, 34; sport fishing 33, 54–55, 59. *See also* fishing industry (commercial capture), sport fishing
Hodgins, Malcolm 59
Hoskins, John 9, 29
Humphrey, Johnny 33, 59
Huu–ay–aht nation: spearing basking sharks 44

Irish fishery 24, 25

Jamie's Whaling Station 63
Japanese fishery 25
Jordan, David Starr 55, 65

Kildonan 46
killer whales 17–18

Lamb, Gordie 57

Mackenzie, Alexander 35
Makah nation: whaling ritual 34
marine conservation 69–70
marine ecosystems: human damage to 12
Matthews, Harrison 21, 22
Maxwell, Gavin 13, 24
McDermid, Bill 58
McHugh, J. 39
McIndoe, Bob 50–51
McLuhan, Winston 59
mergansers 37, 48
Moses, John 44
Motherwell, J.A. 26
mud shark 31, 38. *See also* basking sharks

Natural History Society of BC 25
Neah Bay 32, 34
newspaper accounts: cutting blade 50–52; involving fishing industry 10–11, 37–38, 40–42, 46; killing basking sharks 44; natural history

reports 32; photos of basking sharks 79; sport fishing of basking sharks 23, 33, 55, 59; terminology used to describe basking sharks 50–51, 55, 71, 74–78; war on basking sharks 10–11, 42, 47–52, 57
Nootka Sound 9, 31
Norwegian fishery 24–25
Nuu–chah–nulth 30: shark fisheries 31

Oak Bay 58

Pachena Bay 50, 51, 68
Pachena Point 26
Pacific hake 62
Parksville 26, 43, 54–55
Pauly, Daniel 68
Port Alberni 42
Port Renfrew 32, 34
Puget Sound 32

Qualicum Beach 58, 59
Queen Charlotte Islands 30, 34–35, 62
Queen Charlotte Sound 15, 26, 35

rays: in British Columbia 18–19
razor–billed shark slasher 41, 42
Rivers Inlet 10, 11: salmon fishery 38, 40–42

Ruck, Sidney 38

Saanich Inlet 58, 59
salmon canneries: and basking sharks 10, 11
salmon farming: and basking sharks 27
Sarita 44, 56
satellite tracking: of basking sharks 15, 17, 21
scientific publications: mention basking sharks 25–26; descriptions of basking sharks 55
sea lions 10, 37, 38, 48
sea serpents 42–43
seals 48
sharks: in British Columbia 18–19; debate over rarity 32, 34; meat of 38; shark liver oil 24, 43–44, 47
shifting baseline syndrome 68–69
skates: in British Columbia 18
Smithsonian Institution 59–60
Species at Risk Act 66
sport fishing 11, 23, 33, 54–55, 59, 63, 65
Steller's sea cow 42
Strait of Georgia 15, 26, 43, 57–58
Strait of Juan de Fuca 31–32, 34
Suckley, George 31–32

sun shark 38. *See also* basking shark
supersharks 13
Swan, James G. 32, 34

Texada Island 33, 59
Toba Inlet 37
Tolmie, William Fraser 30

UN Food and Agriculture Organization 13

Vancouver Public Aquarium 59

whale shark 13, 15
white shark 13
Wickham, Eric 57
Wilby, C.V. 26
wildlife tourism: and basking sharks 65–66
World Conservation Union (IUCN) 66: Shark Specialist Group 18

Yarrell, William 20

zooplankton 20, 21, 62

Copyright Scott Wallace and Brian Gisborne 2006

All rights reserved. No part of this work may be reproduced or used in any form or by any means — graphic, electronic, or mechanical — without the prior written permission of the publisher. Any request for photocopying or other reprographic copying must be sent in writing to ACCESS Copyright.

TRANSMONTANUS is edited by Terry Glavin. Editorial correspondence should be sent to Transmontanus, PO Box C25, Fernhill Road, Mayne Island, BC V0N 2J0.

New Star Books Ltd.
107 - 3477 Commercial Street   #1517 – 1574 Gulf Road
Vancouver, BC   V5N 4E8   Point Roberts, WA 98281
www.NewStarBooks.com
info@NewStarBooks.com

Cover by Mutasis Design
Photo on cover: British Columbia Archives No. D-02035
Edited by Betsy Nuse
Map by Eric Leinberger
Typesetting by New Star Books
Printed and bound in Canada by Imprimerie Gauvin
Printed on 100% post-consumer recycled paper
First printing, July 2006

Publication of this work is made possible by grants from the Canada Council, the British Columbia Arts Council, and the Department of Canadian Heritage Book Publishing Industry Development Program.

LIBRARY AND ARCHIVES CANADA CATALOGUING IN PUBLICATION

Wallace Scott, 1969–
  Basking sharks : the slaughter of BC's gentle giants / Scott Wallace and Brian Gisborne.

(Transmontanus)
Includes index.
ISBN 1-55420-022-9

  1. Basking shark — British Columbia — Pacific Coast. 2. Basking shark — control — Government policy — Canada — History. 3. Endangered species — British Columbia — Pacific Coast. I. Gisborne, Brian, 1959– II. Title. III. Series.
QL638.95.C37W34 2006   333.95'633   C2006-900568-0